Lecture Notes in Mathematics

Edited by A. Dold and B. Eckmann

Series: Mathematisches Institut der Universität Erlangen-Nürnberg
Advisers: H. Bauer and K. Jacobs

904

Klaus Donner

Extension of Positive Operators and Korovkin Theorems

Springer-Verlag
Berlin Heidelberg New York 1982

Authors
Klaus Donner
Mathematisches Institut, Universität Erlangen-Nürnberg
Bismarckstraße 1 1/2, 8520 Erlangen

AMS Subject Classifications (1980) 40-A-05, 46-A-22, 46-B-30

ISBN 3-540-11183-2 Springer-Verlag Berlin Heidelberg New York
ISBN 0-387-11183-2 Springer-Verlag New York Heidelberg Berlin

Printed in Germany

Printing and binding: Beltz Offsetdruck, Hemsbach/Bergstr.
2141/3140-543210

Contents

Introduction

Examining the various branches of functional analysis, it will be difficult to find a section in which extension problems of linear operators are completely absent. Usually, there are certain properties such as continuity, positivity, compactness, contractivity, or commutativity with given operators that have to be preserved in connection with the extension of the linear operator in question. For linear functionals, extension problems can often be satisfactorily settled using the Hahn-Banach theorem and its various consequences. Extending linear operators, however, turns out to be a troublesome enterprise.

Analysing the proofs of the known extension theorems for linear mappings we come up with two principal arguments:

1) Continuous linear operators defined on dense subspaces possess a continuous linear extension to the whole space.
2) If H is a linear subspace of an arbitrary real vector space E and F is a Dedekind complete vector lattice, then a linear operator $T : H \to F$ dominated by a sublinear mapping $P : E \to F$ can be extended to a linear operator $\tilde{T}$ defined on E under the domination of P. (The vector-valued version of the Hahn-Banach theorem).

   It is not intended to offer a survey of the numerous results based on these two theorems.

On the other hand, it soon becomes evident that the majority of the extension problems for linear operators resists a direct solution via the methods formulated in (1) and (2). Moreover, several counterexamples (e.g. those concerning the non-existence of certain norm-preserving extensions) are indicating that the available equipment is seriously deficient.

Since the fundamental work of Lindenstrauss (see [45],[46]) mathematical research on the extension of linear operators is indeed dominated by efforts in attaining individual results (e.g. on the existence of certain projections).

The author hopes to interrupt this tradition presenting a complete theory of positive and norm-preserving positive extensions for linear operators in L^p-spaces. In addition, several extension theorems are proved that are also applicable in non-classical Banach lattices.

In Section 2, we shall deduce vector-valued generalizations of extension theorems that are due to Hörmander [32] and Anger/Lembcke [2] for linear forms. The main aspect of these theorems exceeding the Hahn-Banach theorem consists of the fact that the sublinear functionals or mappings do no longer attain their values in $\mathbb{R}$ or a Dedekind complete vector lattice but map into $\mathbb{R} \cup \{+\infty\}$ or into an abstract cone. In general, these cones cannot be imbedded into a vector lattice and possess "infinitely big" elements. A short exposition of some fundamental properties of abstract cones, repeatedly used later, is presented in Section 1.

To get an idea in which way the concept opens a successful approach to extension problems, consider a positive linear operator T from a linear subspace H of $L^p(\mu)$, $1 \leq p < \infty$, into itself, where μ is a σ-finite positive measure. For every positive linear extension $T_o : L^p(\mu) \to L^p(\mu)$ of T we obtain

$$T_o f \leq Pf := \inf\{Th : h \in H, h \geq f\} \quad (f \in L^p(\mu)),$$

the infimum being formed in the cone of all μ-measurable, numerical functions possessing a lower bound in $L^p(\mu)$. (Here we identify μ-almost everywhere coinciding functions and set $\inf \emptyset = +\infty$). In general, we cannot expect Pf to be an element of $L^p(\mu)$ for each $f \in L^p(\mu)$. In fact, this will only be true when each function $f \in L^p(\mu)$ is dominated by some element $h \in H$, an assumption on H that is obviously too strong.

Thus, in cases of practical interest, we have to admit that the function Pf attains the value $+\infty$ on a set of positive measure.
So far, there is nothing new about this approach. Actually, sublinear mappings of this type have already been discussed in literature ([79], [54],[58]). Extension theorems, however, have been obtained only when the set of all elements f in the domain of P satisfying $Pf < \infty$, has non-empty interior with respect to the finest locally convex topology. This assumption, however, is a severe obstacle to applications.
The extension theorems proved here will not suffer from this handicap. On the other hand, things will not work without any regularity conditions (such as lower semicontinuity) on the sublinear mapping P. It must be approximated pointwise from below by locally defined concave mappings dominated by P. Although we shall check this condition in several examples (see 2.13.3), it may be difficult to be verified in various other applications. Moreover, we shall show that the extension technique developed in Section 2 cannot be used to construct norm-preserving extensions of linear operators in non-AM-spaces.

To procede further, we therefore open a new approach in Section 3. If E and F are normed spaces, then every continuous linear operator T from E into the topological dual F' of F induces a continuous bilinear form b_T on $E \times F$. This, in turn, corresponds to a linear form $T^{\otimes}$ on $E \otimes F$ continuous with respect to the projective norm q given by

$$T^{\otimes}(e \otimes f) = b_T(e,f) = Te(f) \text{ for all } e \in E,\ f \in F.$$

Note that $\|T\| \leq 1$ if and only if $b_T(e,f) \leq \|e\| \cdot \|f\|$ or, equivalently, $T^{\otimes} \leq q$. We thus seem to have reduced the problem of norm-preserving linear extensions to the classical Hahn-Banach theorem for linear functionals. Unfortunately, this conclusion is false. To expose the fallacies note first that the operator T to be extended is defined on some subspace $H \subset E$. Let us replace the functional $(e,f) \to \|e\| \cdot \|f\|$

by an arbitrary bisublinear functional $p : E \times F \to \mathbb{R}$. Passing to the tensor product $E \otimes F$ we obtain

$$p^{\otimes}(t) = \inf\{\sum_i p(e_i,f_i) : e_i \in E,\ f_i \in F,\ \sum_i e_i \otimes f_i = t\}, \quad t \in E \otimes F,$$

which will define a sublinear form $p^{\otimes}$ or the constant $-\infty$.
At any rate a bilinear form $b : E \otimes F \to \mathbb{R}$ is dominated by p if and only if $b^{\otimes}$ is dominated by $p^{\otimes}$. Thus we first must make sure that there exists at least one linear operator $S : E \to F'$ such that b_S is dominated by p. If $p(e,f) = \|e\| \cdot \|f\|$, then this condition is clearly satisfied by the zero-operator.
The second obstacle is not so evident. Once we know that

$$Th(f) \leq p(h,f) \quad \text{for all } h \in H,\ f \in F,$$

is it true that $T^{\otimes}(h \otimes f) \leq p^{\otimes}(h \otimes f)$ for all $h \in H$, $f \in F$?
This is trivial when $H = E$, but there are counterexamples for $H \neq E$.
This observation led us to the introduction of so-called subbilinear forms, which, by definition, satisfy the equation

$$p^{\otimes}(e \otimes f) = p(e,f) \quad \text{for all } e \in E,\ f \in F. \quad \text{(see Definiton 3.6).}$$

For $p(e,f) = \|e\| \cdot \|f\|$, p is such a subbilinear form.
There is, however, a third fallacy in the arguments sketched above. Suppose that p actually is a subbilinear form. Then the inequality $Th(f) \leq p(h,f)$ obviously implies $T^{\otimes}(h \otimes f) \leq p^{\otimes}(h \otimes f)$ for all $h \in H$, $f \in F$. But what we need is the domination of $T^{\otimes}$ by $p^{\otimes}$ on all of $H \otimes F$ not only on the elementary tensors $h \otimes f$. When forming $p^{\otimes}(t)$ for $t \in H \otimes F$, however, we have to take into account all the representations $\sum_i e_i \otimes f_i = t$, $e_i \in E$, $f_i \in F$, just choosing $e_i \in H$ will not do! (This is the reason why counterexamples for the existence of norm-preserving extensions can be found even in finite-dimensional spaces).

There is only one possibility to solve extension problems by tensor product methods. We have to be able to compute $p^{\otimes}(t)$ for arbitrary

$t \in E \otimes F$, not only for elementary tensors! This can be successfully done for the bisublinear functional belonging to norm-preserving positive extensions of positive linear operators whenever $F = G'$, where (E,G) is one of the following pairs of Banach lattices:

1) E an arbitrary Banach lattice, G an L^1-space (see 4.3),
2) E an L^p-space, G an L^q-space, where $q \leq p$; $p,q \in [1,\infty[$, (see 5.5),
3) E an AM-space, there exists a positive contractive projection from G'' onto G and G' possesses a topological orthogonal system, (see 5.10).

For such pairs of Banach lattices, called adapted pairs, we obtain a complete solution of the positive and of the positive norm-preserving extension problem. In addition, starting with a positive continuous operator $T: H \to G$, instead of $T: H \to F' = G''$, we end up with an extension $T_o: E \to G$, instead of $T_o: E \to G''$. Given $M \geq 0$, $\|T\| \leq M$, T has a positive linear extension T_o with the norm $\|T_o\| \leq M$ if and only if

$$\|\bigvee_{i\in I}(Th_i)^+\| \leq M\|\bigvee_{i\in I} h_i^+\| \quad \text{for every finite family } (h_i) \text{ in } H,$$

where $\bigvee$ denotes the supremum operator (see 4.4). From this condition several extension and projection theorems in classical Banach lattices can be easily deduced. One of the most remarkable consequences is the following strikingly simple solution of the positive extension problem for adapted pairs (E,G) of Banach lattices (see 4.7):

A positive linear operator $T: H \to G$ possesses a positive linear extension $T_o: E \to G$ if and only if $T(A)$ is bounded from above in G for each subset $A \subset H$ bounded from above in E.

By a counterexample we show that the stated extension theorems are specific for classical Banach lattices (see 4.8).

For the treatment of the applications in Sections 6 to 8 we require detailed information about the set $\mathcal{E}_T$ of all positive linear extensions $\tilde{T}$ of the positive operator $T: H \to G$. In particular, we have to determine the elements $e \in E$ for which the set $\{\tilde{T}e : \tilde{T} \in \mathcal{E}_T\}$ is a singleton, i.e. for which all positive linear extensions $\tilde{T}$ of T coincide at e. This problem is the connecting link of all sections. It is solved for adapted pairs of Banach lattices in Sections 4 and 5. The resulting description of $\{\tilde{T}e : \tilde{T} \in E_T\}$ is extensively applied in the following sections.

While until the end of Section 5 examples and applications have been interspersed occasionally, Section 6 to 8 are concerned with convergence theorems for nets and of positive linear operators. Our starting point is the theorem of Korovkin [39]. It states that for a given sequence $(T_n)_{n\in\mathbb{N}}$ of positive linear operatos on $C([a,b])$ into itself, $a,b \in \mathbb{R}$, $a < b$, $(T_n f)_{n\in\mathbb{N}}$ converges uniformly to f for <u>each</u> $f \in C([a,b])$ provided that

$$\lim_{n\to\infty} T_n(id^j) = id^j \quad \text{for } j = 0,1,2 \quad \text{(uniformly)},$$

where id denotes the identity mapping on $[a,b]$. In addition, Korovkin proved that a minimal "test set", i.e. a minimal set of functions that replaces the set $\{id^0, id, id^2\}$ is a Chebyshev triple. His results have been generalized to arbitrary compact spaces instead of $[a,b]$ by various mathematicians, most notably by Šaškin [63],[64].

Korovkin theorems are most naturally treated within the setting of topological vector lattices or, in the normed case, within the framework of Banach lattices. In fact, such a general investigation is not only illuminating from the theoretical point of view but also covers new applications in spaces of integrable functions and spaces of continuous functions vanishing at infinity. Consider two real Banach lattices E and F, a linear subspace H of E usually called "test space" and a class $\mathcal{T}$ of nets of positive linear operators of E into F. Given a vector lattice homomorphism $S : E \to F$, the <u>Korovkin closure</u> or shadow $Kor_{\mathcal{T},S}(H)$ of H with respect to $\mathcal{T}$ and S is the set of all elements $e \in E$ that satisfy the following condition:

For each net $(T_i) \in \mathcal{T}$, $(T_i e)$ converges to Se, provided that $(T_i h)$ converges to Sh for all $h \in H$.

A test space H such that $\mathrm{Kor}_{\mathcal{T},S}(H) = E$ is called a Korovkin space (with respect to $\mathcal{T}$ and S). If $E = F := C([a,b])$ and if Π_2 denotes the linear subspace of all polynomials of degree at most 2, then Korovkin's classical theorem states that Π_2 is a Korovkin space in $C([a,b])$ with respect to the set P^* of all sequences of positive linear operators on E. In fact, Π_2 is also a Korovkin space with respect to the class P of all nets of positive linear operators on E.

To point out the connection with the extension problem of linear operators let E and F be arbitrary Banach lattices. In order to show that a given element $e \in E$ does not belong to $\mathrm{Kor}_{P,S}(H)$ it is necessary to construct a net $(T_i)_{i \in I}$ in P such that $\lim_{i \in I} T_i h = Sh$ for all $h \in H$ but $(T_i e)_{i \in I}$ does not converge to Se. If the index set I contains only a single element this amounts to the construction of a positive linear extension T_i of the restriction $S|_H$ such that $T_i e \neq Se$. For an arbitrary index set there are still strong connections between the determination of Korovkin closures and the extension of positive, linear operators. Indeed, in many cases the former problem can be reduced to the latter.

A characterization of $\mathrm{Kor}_{P,S}(H)$ for arbitrary Banach lattices E and F has been given in [20],[21]. It carries over to $\mathrm{Kor}_{P^*,S}(H)$, whenever H possesses a countable algebraic basis. Moreover, the description remains valid for locally convex vector lattices E and F. Examining these results in more detail we find out that we persued the wrong trace. To see this, note first that, in Korovkin's classical theorem, sequences $(T_n) \in P^*$ satisfying $\lim_{n\to\infty} T_n h = h$ for all $h \in \Pi_2$ are automatically equicontinuous. If, for arbitrary Banach lattices E and F, P_e and P_e^* denote the classes of all equicontinuous nets and sequences in P and P^*, respectively, the Banach Steinhaus theorem shows that $\mathrm{Kor}_{P_e^*,S}(H) = \mathrm{Kor}_{P^*,S}(H)$ provided that $\mathrm{Kor}_{P^*,S}(H)$ is of second category in E. Since we are mainly interested in cases where Korovkin closures are as big as possible, we had better characterize $\mathrm{Kor}_{P_e^*,S}(H)$ or, more generally, $\mathrm{Kor}_{P_e,S}(H)$ instead of $\mathrm{Kor}_{P^*,S}(H)$ and $\mathrm{Kor}_{P,S}(H)$.

In general, it is not difficult to give sufficient conditions for an element $e \in E$ to belong to $\mathrm{Kor}_{P_e,S}(H)$, several are listed in Section 6 to 8. In order to provide necessity proofs, however, we have to construct particular equicontinuous nets of positive linear operators from E into F. This can be easily done for AM-spaces (see [9],[75], [28]) as well as for ℓ^p-spaces (see [38],[75],[29]), but for L^p-spaces even <u>finite</u> dimensional Korovkin spaces have not yet been characterized.

To fill this gap we first present in Section 6 various descriptions of $\mathrm{Kor}_{P_e,S}(H)$ generally valid for adapted pairs (E,F) of Banach lattices. Since the characterization of Korovkin closures in $C_o(X)$, the space of all continuous real-valued functions on X vanishing at infinity, where X is some locally compact set, is crucial for the L^p-theory, a brief survey of the results for $C_o(X)$ is presented in the first part of Section 7, following [9]. For a linear subspace $H \subset C_o(X)$ of finite dimension $n \in \mathbb{N}$ to be a Korovkin space in $C_o(X)$ the following condition is necessary and sufficient (see 7.24):

For each $x \in X$ and every choice of $n+1$ non-negative real numbers $\alpha_1,\ldots,\alpha_{n+1}$ and points $x_1,\ldots,x_{n+1} \in X$ such that

$$\sum_{i=1}^{n+1} \alpha_i h(x_i) = h(x) \quad \text{for all } h \in H$$

it follows that $\sum_{i=1}^{n+1} \alpha_i = 1$ and $x_i = x$ whenever $\alpha_i \neq 0$.

The rest of Section 7 is concerned with the characterization of Korovkin closures and Korovkin spaces in L^p-spaces, $1 \leq p < \infty$. More precisely, consider a positive Radon measure μ on a locally compact, σ-compact space X and let S denote the identity on $E = F = L^p(\mu)$. For simplicity, assume that $H \subset E$ is finite dimensional with basis

$\{h_1,\ldots,h_n\}$ and that X is second countable. If $\mathscr{L}^p(\mu)$ denotes the linear space of all p^{th} power μ-integrable real-valued functions on X, choose a function $h_i \in \mathscr{L}^p(\mu)$ in the equivalence class h_i for each $i \in \{1,\ldots,n\}$ and let H denote the linear subspace of $\mathscr{L}^p(\mu)$ generated by $h_1,\ldots,h_n$. The subspace $H \subset L^p(\mu)$ is a Korovkin space in $L^p(\mu)$ if and only if there exists a μ-negligible set $N \subset X$ such that for each $x \in X \setminus N$ the point evaluation ε_x at x is the only positive linear functional on $\mathscr{L}^p(\mu)$ of the form $\sum_{i=1}^{n+1} \alpha_i \varepsilon_{x_i}$, $\alpha_i \geq 0$, $x_i \in X \setminus N$, satisfying

$$\sum_{i=1}^{n+1} \alpha_i h(x_i) = h(x) \quad \text{for all } h \in H \quad \text{(see 7.27).}$$

In Section 8, for an arbitrary vector lattice homomorphism S handy characterizations of $\mathrm{Kor}_{P_e,S}(H)$ are deduced from the results in Section 6. To do this we introduce the notion of S-essential sets, which behave similar to the sets carried by a measure. If the vector lattice homomorphism S is defined on $C_o(X)$, where X is a locally compact space, a function $f \in C_o(X)$ belongs to $\mathrm{Kor}_{P_e,S}(H)$ if and only if the set of all points $x \in X$ for which the Dirac measure ε_x at x is the only positive finite measure on X satisfying $\mu(h) = h(x)$ for all $h \in H$, is an S-essential set (see 8.3). When S is the natural imbedding of a space of continuous functions into an L^p-space the results of Berens-Lorentz [11] can be immediately derived from this characterization.

The author wishes to express his gratitude to all who, by their friendly support and co-operation have contributed to the present work, in particular to Professor Dr. H. Berens and Professor Dr. H. Bauer. While several useful hints in the sections dealing with the extension of positive operators are due to Dr. C. Portenier and Dr. B. Anger, the detailed discussions with Dr. S. Papadopoulou have stimulated the work on Korovkin theorems. Finally, I thank Mrs. E. Schöpf who carefully typed the manuscript.

K. Donner

Notations

If f is a mapping from a set M into a set N and K is a subset of M, then $f|_K$ denotes the restriction of f to K.

$\mathbb{R}_+$ is the set of all non-negative, real numbers, $\mathbb{R}_+^* := \mathbb{R}_+ \setminus \{0\}$, $\mathbb{R}_\infty := \mathbb{R} \cup \{\infty\}$.

If M is an arbitrary set, a function $f : M \to \mathbb{R} \cup \{-\infty, +\infty\}$ is called a _numerical_ function in contrast to _real_ functions $g : M \to \mathbb{R}$.

By a _vector space_ we always mean a _real_ vector space.

The notion of an _operator_ is reserved for _linear_ operators, only. The extensively used terminology of the theory of ordered vector spaces and vector lattices is adopted from [66]. Deferring from the convention used there, however, topological vector lattices are _not_ automatically _Hausdorff_.

In the context of a vector lattice V the notions of _orthogonality_ or _disjointness_ of two elements $x, y \in V$ mean that $\inf(|x|, |y|) = 0$. A Banach lattice E has _p-additive_ norm for $p \in [1, \infty[$, if

$$\|x\|^p + \|y\|^p = \|x + y\|^p$$

for any two elements $x, y \in E$ satisfying $\inf(|x|, |y|) = 0$.

Classical Banach lattices are AM-spaces or Banach lattices possessing p-additive norm for some $p \in [1, \infty[$.

Finally, the symbol ∎ marks the end of a proof.

A list of all symbols used can be found on page 174.

1. Cone embeddings for vector lattices

The classical Hahn-Banach theorem states that a sublinear functional $p: E \to \mathbb{R}$, where E denotes a (real) vector space, is the pointwise supremum of the linear forms on E dominated by p. On the other hand, suprema of arbitrary sets of linear forms may attain the value $+\infty$. It is, in fact, possible and, in view of the applications, also recommendable to reformulate the Hahn-Banach theorem for sublinear functionals $p: E \to \mathbb{R}_\infty$, which are l.s.c. for the finest locally convex topology (see [2], [3]).

When extending operators from a linear subspace $H \subset E$ into some Dedekind complete vector lattice F under the domination by a sublinear mapping p the restriction that the range of p should be contained in F proves even more unsatisfactory for the applications. In fact, we often encounter sublinear mappings that attain values interpretable as suprema of increasing families in F formed in some bigger lattice in which F can be imbedded. To be more precise we need some notations:

1.1 Definitions: Let $(C,+)$ be a commutative semi-group with unit 0 endowed with an exterior operation $\odot : \mathbb{R}_+ \times C \to C$ satisfying the following conditions:

$\lambda\odot(c_1 + c_2) = \lambda\odot c_1 + \lambda\odot c_2$ for all $\lambda \in \mathbb{R}_+, c_1, c_2 \in C$,

$(\lambda_1 + \lambda_2)\odot c = \lambda_1\odot c + \lambda_2\odot c$ for all $\lambda_1, \lambda_2 \in \mathbb{R}_+$, $c \in C$,

$\lambda\odot(\mu\odot c) = (\lambda\mu)\odot c$ for all $\lambda, \mu \in \mathbb{R}_+$, $c \in C$,

$1\odot c = c$ and $0\odot c = 0$ for all $c \in C$.

Then $(C, +, \odot)$ will be called a cone.

If C_o denotes the group of invertible elements in C, we have $\lambda\odot c \in C_o$ for all $c \in C_o$, $\lambda \in \mathbb{R}_+$ since

$\lambda \odot c + \lambda \odot (-c) = \lambda \odot (c + (-c)) = \lambda \odot 0 = \lambda \odot (0 \odot 0) = (\lambda \cdot 0) \odot 0 = 0 \odot 0 = 0.$

Let $\boxdot : \mathbb{R} \times C_o \to C_o$ be given by

$$\lambda \boxdot c = \begin{cases} \lambda \odot c, & \text{if } \lambda \geq 0 \\ -\lambda \odot (-c), & \text{if } \lambda < 0. \end{cases}$$

Then it is easy to check that $(C_o, +|_{C_o \times C_o}, \boxdot)$ is a vector space.

In order to avoid unnecessary formalism we shall write λc instead of $\lambda \odot c$ or $\lambda \boxdot c$ in the sequel and we briefly refer to "the cone C" rather than to "the cone $(C,+,\odot)$".

Given an ordering $\leq$ on a cone C we call $(C,\leq)$ an <u>ordered cone</u>, if, for all $x,y,z \in C, \lambda \in \mathbb{R}_+$

$$x + z \leq y + z \quad \text{and} \quad \lambda x \leq \lambda y \quad \text{whenever} \quad x \leq y.$$

$(C,\leq)$ is said to be a <u>lattice cone</u>, if the supremum $x \vee y$ and the infimum $x \wedge y$ exist in C for any two elements $x,y \in C$. In this case, we have a <u>positive part</u> $x^+ := x \vee 0$ for each $x \in C$.

A lattice cone $(C,\leq)$ is <u>Dedekind complete</u>, if every nonempty subset $A \subset C$ bounded from above or below, respectively, has a least upper (or lower) bound sup A (resp. inf A). Again, by abuse of language, we give up the pair notation $(C,\leq)$ speaking only of "the ordered cone C".

Given a vector lattice F a Dedekind complete lattice cone C is called <u>imbedding cone</u> for F, if $F = C_o$ with coinciding algebraic and order structures on C_o and if

(IC1) $c = \sup\{x \in F : x \leq c\}$ for every $c \in C$

(IC2) $x + f \wedge y = (x + f) \wedge (x + y)$ whenever $x,y \in C, f \in F$.

Furthermore, we shall say that C is a <u>tight</u> imbedding cone provided that

$$\{c \in C : \exists\, x \in C_o : c \leq x\} \subset C_o .$$

For a tight imbedding cone C the vector space C_o is clearly Dedekind complete with coinciding suprema and infima in C_o and C, respectively, for bounded subsets $A \subset C_o$. Moreover, if $x \in C_o$, then $x \wedge 0 \in C_o$, hence we can define the negative part $x^- := -(x \wedge 0)$.

1.2 Examples:

a) Each Dedekind complete vector lattice F is a tight imbedding cone for itself.

b) $\mathbb{R}_\infty$ is tight imbedding cone for $\mathbb{R}$, if we set $0 \cdot \infty := 0$.

c) If S denotes the cone of all l.s.c. $\mathbb{R}_\infty$-valued functions on $[0,1]$ endowed with pointwise operations and order, then S is an imbedding cone for $C([0,1])$ which is not tight.

d) Given a σ-finite measure space $(\Omega, \mathcal{A}, \mu)$ and real number $p \in [1,\infty]$, let M be the set of all $\mathcal{A}$-measurable numerical functions on Ω and $N := \{f \in M : f = 0 \ \mu\text{-a.e.}\}$.

It is well-known that the vector lattice $\mathcal{L}^p(\mu)$ of all $\mathcal{A}$-measurable real-valued functions f on Ω satisfying $\int |f|^p d\mu < \infty$, if $p \neq \infty$, $\mu(\{x \in \Omega : |f(x)| > r\}) = 0$ for some $r \in \mathbb{R}_+$, if $p = \infty$, becomes a complete locally convex vector lattice under the $\mathcal{L}^p$-semi-norm $f \to (\int |f|^p d\mu)^{1/p}$ for $p \neq \infty$ and $f \to \inf\{r \in \mathbb{R}_+ : \mu(\{x \in \Omega : |f(x)| > r\}) = 0\}$ for $p = \infty$, respectively. Furthermore, endowed with the quotient structure, $L^p(\mu) := \mathcal{L}^p(\mu)/_{N \cap \mathcal{L}^p(\mu)}$ is a Banach lattice.

In search for suitable imbedding cones, consider the set $M_p := \{f \in M : \exists \ g \in \mathcal{L}^p(\mu) : g \leq f\}$.

With operations and order defined pointwise, where again we use the convention $0 \cdot \infty := 0$, M_p is a lattice cone.

The same holds for the quotient cone $C_p := M_p/_{M_p \cap N}$ under the induced quotient structure. In contrast to M_p, however, C_p is in fact Dedekind complete (see [53], Ch. 4, § 23, Ex. 3.3, iv), and, since $N \cap \mathcal{L}^p(\mu) \subset N \cap M_p$, we may identify $L^p(\mu)$ with the vector lattice of all invertible elements of C_p. It is easy to check that C_p is a tight imbedding cone for $L^p(\mu)$.

e) Let F be a Dedekind complete Banach lattice with order continuous norm. Then F is a vector lattice ideal in its bidual F'' (cf. [66], Ch. II, 5.10). If C denotes the set of all $\varphi \in F''$ which are suprema in F'' of subsets $A \subset F$, then C is a tight imbedding cone for F.

1.3 Lemma: If C is a tight imbedding cone for a Dedekind complete vector lattice F, the following equalities hold for each $a,b \in C, f \in F$.

i) $a + f = a \vee f + a \wedge f$.

In particular, $a = a + 0 = a \vee 0 + a \wedge 0 = a^+ - a^-$.

ii) $a \vee (f \wedge b) = (a \vee f) \wedge (a \vee b)$

iii) $f \wedge (a \vee b) = (f \wedge a) \vee (f \wedge b)$.

Moreover, if $D \subset C$ is bounded from below, then

iv) $a + \inf D = \inf_{d \in D}(a + d)$.

Proof: i) Since $f - a \wedge f \geq 0$ and $a - a \wedge f \geq 0$, we obtain $a + f - a \wedge f \geq a$ and $a + f - a \wedge f \geq f$, which implies $a + f - a \wedge f \geq a \vee f$. Conversely, for each $g \in F$ such that $g \leq a + f$ it follows that

$$g = (g - f) + f = (g - f) \wedge f + (g - f) \vee f \leq a \wedge f + a \vee f.$$

Hence, $a + f = \sup\{g \in F : g \leq a + f\} \leq a \wedge f + a \vee f$.

ii) Using i) and property IC2 we deduce

$$a \vee (f \wedge b) = a + f \wedge b - a \wedge f \wedge b = (a + f) \wedge (a + b) - a \wedge f \wedge b.$$

If we set $x := a + f$, $y := a + b$, $g := a \wedge f \wedge b$, the inequalities $x \geq (x - g) \wedge (y - g) + g$, $y \geq (x - g) \wedge (y - g) + g$ imply that $x \wedge y \geq (x - g) \wedge (y - g) + g$. Furthermore, since $a \geq g$ and $b \geq g$, we conclude $a + b - g \geq a$ and $a + b - g \geq b$, hence $y - g \geq a \vee b$. Therefore

$$a \vee (f \wedge b) = x \wedge y - g \geq (x - g) \wedge (y - g) \geq (a + f - a \wedge f) \wedge (a \vee b) = \\ = (a \vee f) \wedge (a \vee b).$$

The converse inequality is trivial.

iii) Repeated application of (ii) yields

$$(f \wedge a) \vee (f \wedge b) = (f \vee (f \wedge a)) \wedge (b \vee (f \wedge a))$$

$$= f \wedge (f \vee a) \wedge (b \vee f) \wedge (b \vee a) = f \wedge (b \vee a).$$

iv) Since $a + \inf D \leq a + d$ for all $d \in D$ we have $a + \inf D \leq \inf_{d \in D}(a + d)$.

Conversely, given $f \in F$ such that $f \leq a + d$ for each $d \in D$, we obtain

$$f + d^- = (f \wedge (a + d)) + d^- = (f + d^-) \wedge (a + d + d^-) = (f + d^-) \wedge (a + d^+)$$

$$\leq (f + d^- + d^+) \wedge (a + d^+) = (f + d^-) \wedge a + d^+ \quad \text{for each } d \in D.$$

Consequently, $f - (f + d_o^-) \wedge a \leq f - (f + d^-) \wedge a \leq d$ for all $d \in D$, where $d_o := \inf D$. We thus conclude

$$f = (f + d_o^-) \wedge a + (f - (f + d_o^-) \wedge a) \leq a + \inf D.$$

Using (IC1), it follows that

$$\inf_{d \in D}(a + d) = \sup\{f \in F : f \leq \inf_{d \in D}(a + d)\} \leq a + \inf D. \quad \blacksquare$$

We shall frequently use imbedding cones. Hence it is important to ensure the existence of imbedding cones for Dedekind complete vector lattices. This is done in the following

<u>1.4 Theorem</u>: Given a Dedekind complete vector lattice F there exists a tight imbedding cone C with the following properties

i) C has a biggest element.

ii) For any two non-empty subsets $A, B \subset C$ satisfying $\sup A = \sup B$ the equality $\sup_{a \in A}(a \wedge f) = \sup_{b \in B}(b \wedge f)$

holds for each $f \in F$.

Furthermore, C is a distributive lattice and is uniquely determined up to isomorphisms.

If C_1 is an arbitrary tight imbedding cone of F the mapping $J : C_1 \to C$ given by

$$J(x) = \sup\{f \in F : f \leq x \text{ in } C_1\}$$

where the supremum is formed in C, defines a lattice cone monomorphism. In particular, C_1 is a distributive lattice.

Proof: 1st step: Construction of C.

Consider the system $\mathcal{A}$ of all non-empty, upward directed subsets of F. For $A,B \in \mathcal{A}$ we define the equivalence relation

$$A \sim B \Leftrightarrow \forall f \in F : \sup_{a\in A}(a \wedge f) = \sup_{b\in B}(b \wedge f).$$

It is easy to check that the sum $A + B = \{a + b : a \in A,\ b \in B\}$ and the product $\lambda A = \{\lambda a : a \in A\}$ ($A,B \in \mathcal{A},\ \lambda \in \mathbb{R}_+$) are compatible with the equivalence realtion $\sim$. Endowed with the quotient operations $C := {}^{\mathcal{A}}/_{\sim}$ is a cone. Assigning the equivalence class of $\{f\} \in \mathcal{A}$ to each $f \in F$ yields a vector space monomorphism Λ of F into the linear space C_o of all (additively) invertible elements of C. In fact, Λ is an isomorphism, since for each $A \in \mathcal{A}$ with invertible equivalence class $[A]$ there is $B \in \mathcal{A}$ such that $A + B \sim \{0\}$, i.e. $\sup\{a + b : a \in A,\ b \in B\} = 0$. Hence $a + b \leq 0$, which implies $a \leq -b$ for all $a \in A$, $b \in B$. F being Dedekind complete $f := \sup A$ exists in F and $\{f\} \sim A$. Obviously, $\Lambda(f) = [A]$. If we set

$$A \wedge B = \{a \wedge b : a \in A,\ b \in B\},$$
$$A \vee B = \{a \vee b : a \in A,\ b \in B\}$$

for $A,B \in \mathcal{A}$, the operations $\wedge$, $\vee$ are compatible with the equivalence relation $\sim$. Moreover, each of the following equalities implies the next and vice versa:

$$\sup_{a\in A,b\in B}(a \wedge b \wedge f) = \sup_{a\in A}(a \wedge f)$$

$$\sup_{a\in A}(a \wedge f) \wedge \sup_{b\in B}(b \wedge f) = \sup_{a\in A}(a \wedge f)$$

$$\sup_{b\in B}(b \wedge f) \geq \sup_{a\in A}(a \wedge f)$$

$$\sup_{a\in A}(a \wedge f) \vee \sup_{b\in B}(b \wedge f) = \sup_{b\in B}(b \wedge f)$$

$$\sup_{(a,b)\in A\times B} ((a \vee b) \wedge f) = \sup_{(a,b)\in A\times B} ((a \wedge f) \vee (b \wedge f)) = \sup_{b\in B}(b \wedge f).$$

Hence, if we define

$$[A] \leq [B] :\Leftrightarrow \sup_{a\in A}(a \wedge f) \leq \sup_{b\in B}(b \wedge f)$$

for $A,B \in \mathcal{A}$, then C becomes a lattice under this ordering and

$$[A] \wedge [B] = [A \wedge B], \quad [A] \vee [B] = [A \vee B].$$

As an immediate consequence we obtain the distributivity laws for $\wedge, \vee$. Furthermore, since the inclusion $(A \vee A') + B \subset (A+B) \vee (A'+B)$ holds for each $A,A',B \in \mathcal{A}$, it follows that

$$([A] \vee [A']) + [B] = [(A \vee A') + B] \leq$$
$$\leq [(A+B) \vee (A'+B)] = ([A]+[B]) \vee ([A']+[B]).$$

Conversely,

$$[A]+[B] \leq ([A] \vee [A']) + [B] \text{ and } [A']+[B] \leq ([A] \vee [A']) + [B],$$

hence

$$([A] \vee [A']) + [B] = ([A]+[B]) \vee ([A']+[B]).$$

In particular, if $[A] \leq [A']$, then $[A']+[B] = ([A]+[B]) \vee ([A']+[B])$, which implies that $[A]+[B] \leq [A']+[B]$. Since, obviously, $\lambda[A] \leq \lambda[A']$ for all $\lambda \in \mathbb{R}_+$, $A,A' \in \mathcal{A}$ satisfying $[A] \leq [A']$, C is a lattice cone with biggest element $[F]$ and the mapping $\Lambda : F \to C_o$ defined above is a vector lattice isomorphism. The proof of the Dedekind completeness of C is straightforward, hence we omit the details.

Finally, it follows from the equality $[A] = \sup_{a\in A}[\{a\}]$ and the relation $[\{a\}] \in \Lambda(F)$, valid for each $a \in A$, that $[A] = \sup\{\Lambda(f) : f \in F, \Lambda(f) \leq [A]\}$ for each $a \in \mathcal{A}$, which proves property (IC1) of imbedding cones. In order to show (IC2) let $A,A',B \in \mathcal{A}$ and $a \in A$, $a' \in A'$, $b_1,b_2 \in B$ be given. If $b \in B$ satisfies $b \geq b_1$ and $b \geq b_2$, then

$$(a+b_1) \wedge (a'+b_2) \leq (a+b) \wedge (a'+b) = (a \wedge a') + b \in A \wedge A' + B.$$

From the resulting inequality

$$\inf_{\substack{a\in A, a'\in A'\\ b_1, b_2\in B}} ((a+b_1)\wedge(a'+b_2)\wedge f) \leq \inf_{\substack{a\in A, a'\in A'\\ b\in B}} (((a\wedge a')+b)\wedge f) \qquad (f\in F)$$

we deduce

$$([A]+[B])\wedge([A']+[B]) = [(A+B)\wedge(A'+B)] \leq$$
$$\leq [(A\wedge A')+B] \leq ([A]\wedge[A'])+[B].$$

Since the converse inequality is obvious, we conclude

$$([A]+[B])\wedge([A']+[B]) = ([A]\wedge[A'])+[B],$$

which proves C to be an imbedding cone for $\Lambda(F)$.

It is easy to verify the tightness as well as property (ii) of C.

For the rest of the proof we shall identify F and $\Lambda(F)$.

2nd step:

Given an arbitray tight imbedding cone C_1 of F we put

$$J(x) := [A_x], \text{ where } A_x := \{f\in F : f\leq x \text{ in } C_1\} \text{ for each } x\in C_1.$$

The mapping $J : C_1 \to C$ is injective, since for any two elements $x,y\in C_1$ satisfying $A_x \sim A_y$ we obtain in C_1

$$x = \sup A_x = \sup_{f\in A_x} (\sup_{f'\in A_x} f\wedge f') = \sup_{f\in A_x} \sup_{g\in A_y} f\wedge g$$

using the relation $A_x \sim A_y$, hence

$$x = \sup_{g\in A_y} \sup_{f\in A_x} f\wedge g = \sup_{g\in A_y} \sup_{g'\in A_y} g'\wedge g = y.$$

In order to show the additivity of J consider two elements $x,y\in C_1$.
For each $f\in A_{x+y}$ we have $(f+y^-)\wedge x\in A_x$ and

$$f+y^- = f\wedge(x+y)+y^- = (f+y^-)\wedge(x+y^+) \leq$$
$$\leq (f+y^-+y^+)\wedge(x+y^+) = (f+y^-)\wedge x+y^+,$$

which implies that $f-(f+y^-)\wedge x\in A_y$. Consequently,
$f = (f+y^-)\wedge x+(f-(f+y^-)\wedge x)\in A_x+A_y$.
The inclusion $A_x+A_y \subset A_{x+y}$ being evident, we deduce $A_{x+y} = A_x+A_y$.
From this equality, the additivity of J immediately follows:

$$J(x+y) = [A_{x+y}] = [A_x] + [A_y] = J(x) + J(y).$$

Furthermore, since $A_x \wedge A_y = A_{x\wedge y}$ for $x,y \in C_1$, we obtain

$$J(x \wedge y) = [A_{x\wedge y}] = [A_x \wedge A_y] = [A_x] \wedge [A_y] = J(x) \wedge J(y).$$

Similarly, in order to prove $J(x \vee y) = J(x) \vee J(y)$, it suffices to check $A_{x\vee y} \subset A_x \vee A_y$, the converse inclusion being obvious. This, however, follows immediately from Lemma 1.3, iii, since

$$f = f \wedge (x \vee y) = (f \wedge x) \vee (f \wedge y) \in A_x \vee A_y \quad \text{for each } f \in A_{x\vee y}.$$

It remains to show that the imbedding cone C is uniquely determined up to isomorphisms by the properties (i) and (ii). To this end, for an tight imbedding cone C' of F satisfying (i) and (ii), consider the cone isomorphism $J: C' \to C$ defined above. We claim that J maps C' onto C. Given $A \in \mathcal{A}$, since C' is Dedekind complete and has a biggest element $x := \sup A$ exists <u>in C'</u>. Moreover, from the equality $\sup A_x = x = \sup A$ valid in C', we deduce

$$\sup_{a \in A_x} (a \wedge f) = \sup_{a \in A} (a \wedge f) \quad \text{for each } f \in F$$

using property (ii). Therefore, $A_x \sim A$, which yields $J(x) = [A]$. Consequently, J is a lattice cone isomorphism of C' onto C. ∎

<u>1.5 Definition</u>: If F is a Dedekind complete vector lattice, then the tight imbedding cone satisfying the conditions (i) and (ii) of Theorem 1.4 is called the <u>sup-completion</u> of F, denoted by F_s. The cone monomorphism $J: C_1 \to F_s$ defined for an arbitrary tight imbedding cone C_1 of F according to Theorem 1.3 is said to be the <u>canonical imbedding of C_1 into F_s</u>.

<u>1.6 Examples</u>:

1) The sup-completion of $\mathbb{R}$ is clearly $\mathbb{R}_\infty$.

2) If $(\Omega, \mathcal{A}, \mu)$ is a σ-finite measure space and $p \in [1,\infty]$, the lattice

cone C_p introduced in 1.2,d is the sup-completion of $L^p(\mu)$.

3) Given $p \in [1,\infty]$, the sup-completion of ℓ^p is the cone of all $\mathbb{R}_\infty$-valued sequences, possessing an ℓ^p-minorant.

<u>1.7 Remark</u>: If C is a tight imbedding cone of a vector lattice F then, in general, the extended distributivity relation

$$(+) \qquad \sup_{i \in I}(a_i \wedge f) = (\sup_{i \in I} a_i) \wedge f$$

does not hold for each family $(a_i)_{i \in I} \subset C$ bounded from above and for each $f \in F$. This is true, even if C contains a maximal element. As a counterexample, consider the cone $C = \mathbb{R}^2 \cup \{(\infty,\infty)\}$ with operations and ordering defined pointwise. Then C is clearly a tight imbedding cone for $\mathbb{R}^2$ which is not isomorphic to the sup-completion $\mathbb{R}_\infty \times \mathbb{R}_\infty$ of $\mathbb{R}^2$. Hence, by Theorem 1.4, the equality (+) cannot hold in C.

2. A vector-valued Hahn-Banach Theorem

2.1 Definition: a) Given a vector space E and an ordered cone C, a mapping $p : E \to C$ is called sublinear (resp. superlinear), if

$p(x+y) \leq p(x) + p(x)$ for all $x,y \in E$ (subadditively)

$(p(x+y) \geq p(x) + p(y)$ for all $x,y \in E$, respectively), and

$p(\lambda x) = \lambda p(x)$ for all $\lambda \in \mathbb{R}_+$, $x \in E$ (positive homogenity).

b) Let K be a convex subset of a vector space E and let C be an ordered cone. A mapping $f : K \to C$ is convex, if

$$f(\lambda x + (1-\lambda)y) \leq \lambda f(x) + (1-\lambda) f(y) \quad \text{for all } \lambda \in [0,1],\ x,y \in K.$$

it is called concave, if

$$f(\lambda x + (1-\lambda)y) \geq \lambda \cdot f(x) + (1-\lambda) f(y) \text{ for all } \lambda \in [0,1],\ x,y \in K.$$

2.2 Lemma: Let E be a locally convex vector space, U a convex neighborhood of O in E and let C be a tight imbedding cone for some vector lattice F. If $k : U \to F$ is concave and satisfies $k(0) \leq 0$, then

$$\tilde{k}(x) := \begin{cases} \sup\{\frac{1}{\lambda} k(\lambda x) : \lambda \in \mathbb{R}_+^*,\ \lambda x \in U\} & \text{for } x \neq 0 \\ 0 & \text{for } x = 0 \end{cases}$$

is an element of F for each $x \in E$. $\tilde{k}$ is the smallest positively homogeneous function $q : E \to C$ such that $q|_U \geq k$ and $\tilde{k}$ is superlinear. Moreover, if F is a topological vector lattice and if k is continuous at O, the $\tilde{k}$ is continuous.

Proof: Given $x \in E \setminus \{0\}$, choose $\varepsilon > 0$ such that $[-\varepsilon x, \varepsilon x] \subset U$. If $\lambda \in \mathbb{R}_+^*$, $\lambda > \varepsilon$, is such that $\lambda x \in U$ and if $\alpha := \min(\frac{\varepsilon}{\lambda}, \frac{1}{2}) \in]0,1[$ we conclude from the concavity of k:

$k(\alpha\lambda x) = k(\alpha(\lambda x) + (1-\alpha)\cdot 0) \geq \alpha \cdot k(\lambda x) + (1-\alpha) k(0)$, whence

$$(2.2.1) \qquad \frac{1}{\lambda}k(\lambda x) \leq \frac{1}{\alpha\lambda}\,k(\alpha\lambda x) - \frac{1-\alpha}{\alpha\lambda}k(0).$$

Furthermore, for $\mu \in]0,\varepsilon], \beta := \frac{\varepsilon}{\varepsilon+\mu}$, we obtain

$k(0) = k((\beta\mu + (1-\beta)(-\varepsilon))\cdot x) \geq \beta k(\mu x) + (1-\beta)k(-\varepsilon\cdot x)$, which yields

$$(2.2.2) \quad \frac{1}{\mu}\cdot k(\mu x) \leq \frac{1}{\mu\beta}(k(0) - (1-\beta)k(-\varepsilon x)) \leq -\frac{1-\beta}{\mu\beta}k(-\varepsilon x) = -\frac{1}{\varepsilon}k(-\varepsilon x).$$

Inserting $\mu := \alpha\lambda$ in 2.2.2 and using (2.2.1) we deduce
$\frac{1}{\lambda}k(\lambda x) \leq -\frac{1}{\varepsilon}k(-\varepsilon x) - \frac{1-\alpha}{\alpha\lambda}k(0)$, and, observing the inequality
$\frac{1-\alpha}{\alpha\lambda} \leq \frac{1}{\varepsilon}$,

$$(2.2.3) \qquad \frac{1}{\lambda}k(\lambda x) \leq \frac{1}{\varepsilon}|k(-\varepsilon x)| + \frac{1}{\varepsilon}|k(0)|.$$

Combining (2.2.2) and (2.2.3) we conclude

$$(2.2.4) \qquad \frac{1}{\varkappa}k(\varkappa x) \leq \frac{1}{\varepsilon}|k(-\varepsilon x)| + \frac{1}{\varepsilon}|k(0)|$$

for every $\varkappa \in \mathbb{R}_+^*$ such that $\varkappa x \in U$.
Therefore the set $\{\frac{1}{\varkappa}k(\varkappa x) : \varkappa \in \mathbb{R}_+^*, \varkappa x \in U\}$ is bounded from above in F, which shows that $\tilde{k}(x) \in F$.
Obviously, $\tilde{k}$ is positively homogeneous. In order to prove the superadditivity, let $u,v \in E$ be given.
For $u = 0$ or $v = 0$ the equality $\tilde{k}(u+v) = \tilde{k}(u) + \tilde{k}(v)$ holds. We may therefore assume that $u \neq 0$, $v \neq 0$. If $\lambda,\mu \in \mathbb{R}_+^*$ satisfy $\lambda u \in U$ and $\mu v \in U$ we deduce from the convexity of U that

$$\frac{\mu\lambda}{\mu+\lambda}(u+v) = \frac{\mu}{\mu+\lambda}\lambda u + \frac{\lambda}{\mu+\lambda}\mu v \in U, \text{ hence}$$

$$k\left(\frac{\lambda\mu(u+v)}{\mu+\lambda}\right) \geq \frac{\mu}{\mu+\lambda}\,k(\lambda\mu) + \frac{\lambda}{\mu+\lambda}\,k(\mu v).$$

Consequently,

$$\tilde{k}(u+v) \geq \frac{\mu+\lambda}{\mu\lambda}\,k\left(\frac{\mu\lambda}{\mu+\lambda}(u+v)\right) \geq \frac{1}{\lambda}\,k(\lambda u) + \frac{1}{\mu}\,k(\mu v).$$

Passing to the respective suprema on the right side, we obtain
$\tilde{k}(u+v) \geq \tilde{k}(u) + \tilde{k}(v)$.
Since $\tilde{k}(x) \geq \frac{1}{1}\cdot k(1\cdot x) = k(x)$ for all $x \in U$, $\tilde{k}$ dominates k on U. Finally, consider a positively homogeneous mapping $q : E \to C$ satisfying $q|_U \geq k$.

For each $x \in E$ and $\lambda \in \mathbb{R}_+^*$ such that $\lambda x \in U$ it then follows that

$$\frac{1}{\lambda} k(\lambda x) \leq \frac{1}{\lambda} q(\lambda x) = q(x).$$

Hence $\tilde{k}(x) \leq q(x)$.

Suppose now that F is a (Dedekind complete) topological vector lattice and that $k : U \to F$ is continuous at 0.

Given a solid zero-neighborhood V in F, choose a solid zero neighborhood V' in F satisfying $V' + V' \subset V$ and a solid zero-neighborhood $U' \subset U$ such that

$$k(0) - k(x) \in V' \quad \text{whenever} \quad x \in U'.$$

Applying inequality (2.2.4) for $x \in U'$, $\varepsilon := 1$, we obtain

$$-|k(0)| - |k(x) - k(0)| \leq -|k(x)| \leq k(x) \leq \tilde{k}(x)$$
$$\leq |k(-x)| + |k(0)| \leq 2|k(0)| + |k(-x) - k(0)|.$$

For $\sigma \in]0,1[$ satisfying $2\sigma|k(0)| \in V'$ it follows that

$$\sigma(2|k(0)| + |k(-x) - k(0)|) \in V' + V' \subset V \text{ and}$$
$$\sigma(-|k(0)| - |k(x) - k(0)|) \in V' + V' \subset V$$

using the symmetry of U' and V'.

Since V is solid, we conclude $\sigma\tilde{k}(x) \in V$ for all $x \in U'$ which yields $\tilde{k}(x) \in V$ for all $x \in \sigma U'$.

Therefore $\tilde{k}$ is continuous at 0, and, by the superlinearity of $\tilde{k}$, at any point of E. ∎

<u>2.3 Proposition</u>: Let E be a locally convex vector space, C a tight imbedding cone of a topological vector lattice F and let $p : E \to C$ be a sublinear mapping. Then the following statements are equivalent:

i) There is a continuous operator $T : E \to F$ such that $T \leq p$.

ii) There exists a convex zero-neighborhood U in E and a concave mapping $k : U \to F$ such that k is continuous at 0 and $k(x) \leq p(x)$ for all $x \in U$.

iii) There is a continuous superlinear mapping $s : E \to F$ such that $s \leq p$.

Proof: While (i) ⇒ (ii) is obvious, the implication (ii) ⇒ (iii) follows at once from Lemma 2.2. To prove (iii) ⇒ (i) let $s : E \to F$, $s \leq p$, be a continuous superlinear mapping. For all $x, y \in E$ we then obtain

$s(x) \leq s(x - y) - s(-y) \leq p(x - y) - s(-y)$.

Hence $q(x) = \inf_{y \in E}(-s(-y) + p(x - y))$ exists in F for each $x \in E$. Evidently, q is positively homogeneous. In order to show the subadditivity, let $a, b \in E$ be given. Using Lemma 1.3, iv we deduce

$$
\begin{aligned}
q(a+b) &= \inf_{y \in E}(-s(-y) + p(a+b-y)) \\
&= \inf_{u,v \in E}(-s(-(u+v)) + p(a+b-(u+v))) \\
&\leq \inf_{u,v \in E}(-s(-u) - s(-v) + p(a-u) + p(b-v)) \\
&= \inf_{u \in E} \inf_{v \in E}(-s(-u) + p(a-u) + (-s(-v) + p(b-v))) \\
&= \inf_{u \in E}(-s(-u) + p(a-u) + \inf_{v \in E}(-s(-v) + p(b-v))) \\
&= \inf_{u \in E}(-s(-u) + p(a-u)) + \inf_{v \in E}(-s(-v) + p(b-v)) \\
&= q(a) + q(b).
\end{aligned}
$$

By the vector-valued version of the classical Hahn-Banach theorem (cf. [60], Ch. II, § 2, Prop. 2.1) there exists an operator $T : E \to F$ such that $T \leq q$. Since $q(x) \leq -s(-x)$ for all $x \in E$, q is continuous. Hence (i) follows from the relation $q \leq p$. ■

2.4 Corollaries:

a) (see [2], Thm. 1.8). If E is a locally convex vector space and $p : E \to \mathbb{R}_\infty$ is a sublinear functional, a continuous linear form f on E satisfying $f \leq p$ exists if and only if p is l.s.c. at 0. In fact, if p is l.s.c. at 0 there exists a convex zero-neighborhood

U in E such that $p(x) \geq -1$ for all $x \in U$. The constant function -1 defined on U being concave, the conclusion follows from Proposition 2.3.

b) Consider a sublinear mapping p from a vector space E into some tight imbedding cone of a vector lattice F. An operator $T : E \to F$ satisfying $T \leq p$ exists if and only if there is a concave mapping $k : U \to F$ defined on some convex, absorbing set $U \subset E$ such that $k \leq p|_U$. This is an immediate consequence of Proposition 2.3, endowing E with the finest and F with the coarsest locally convex topology. Since the only zero-neighborhood in F is F itself, k is then clearly continuous.

c) Again, let E be a vector space, C a tight imbedding cone for some topological vector lattice F, and let $p : E \to C$ be sublinear. Suppose that $p(x) \in F$ and $-p(x) = p(-x)$ for some point $x \in E$. If there exists a concave continuous mapping $k : U \to F$ defined on a convex zero-neighborhood $U \subset E$ such that $k(y) \leq p(y)$ for all $y \in U$, then we can find a continuous operator $T : E \to F$ satisfying $T \leq p$ and $T(x) = p(x)$. Indeed, if $T : E \to F$ is any operator such that $T \leq p$, we have $Tx \leq p(x)$ and $-Tx = T(-x) \leq p(-x) = -p(x)$, hence $Tx = p(x)$.

2.5 Remark: If E is a locally convex vector space and C is a tight imbedding cone of a topological vector lattice F, then we can define the lower semi-continuity of a sublinear mapping $p : E \to C$ at 0 as follows:

p is l.s.c. at 0 provided that for each zero-neighborhood V in F there is a zero-neighborhood U in E such that

$$p(x)^- \in V \quad \text{for all } x \in U.$$

In general, however, it is impossible to prove a vector-valued version of the statement in 2.4,a (see counterexample 3.25). - A generalization of 2.4, a to l.s.c. sublinear mappings $p : E \to C$ and operators $T : E \to F$ remains true, if the positive cone F_+ has an interior point.

This can be shown by an obvious modification of the argument in 2.4,a. (cf. also [58]).

2.6 Notations: Let p be a sublinear mapping from a locally convex vector space E into a tight imbedding cone C for some topological vector lattice F. If $A \subset E$, a map $f : A \to C$ satisfying $f(x) \leq p(x)$ for all $x \in A$, is called p-dominated. For $x \in E$ we denote by $K_{p,x}$ the set of all p-dominated, concave mappings $k : x + U \to F$ continuous at x, where U ranges over all convex zero-neighborhoods in E.

Let H be a linear subspace of E and $T : H \to F$ a continuous, p-dominated operator such that the set

$$\{Th + p(x-h) : h \in H\}$$

is bounded from below in C. We then denote by p_T the sublinear mapping from E into C defined by

$$p_T(x) = \inf_{h \in H}(Th + p(x-h)).$$

With these notations we can state

2.7 Lemma: An operator $T_o : E \to F$ is p_T- dominated iff T_o is a p-dominated linear extension of T.

Proof: Since $T(h-h') \leq p(h-h')$ for each $h,h' \in H$, we obtain $Th \leq Th' + p(h-h')$. Consequently, $Th \leq p_T(h)$. The converse inequality being evident, it follows that $Th = p_T(h)$. Furthermore, since $p_T(x) \leq T(0) + p(x-0) = p(x)$ for all $x \in E$, every p_T-dominated operator $T_o : E \to F$ is p-dominated and $T_o h \leq Th$, $T_o(-h) \leq T(-h)$ for each $h \in H$. Therefore T_o must be a linear extension of T.

Conversely, if $T_o : E \to F$ is a p-dominated, linear extension of T, then $T_o(x) - T(h) = T_o(x-h) \leq p(x-h)$ for all $x \in E$, $h \in H$. Hence $T_o(x) \leq \inf_{h \in H}(Th + p(x-h)) = p_T(x)$. ∎

2.8 Theorem: Let p be a sublinear mapping from a locally convex vector space E into a tight imbedding cone of some topological vector lattice F. Given $x \in E$ and $f \in F$, there exists a continuous p-dominated operator $T : E \to F$ satisfying $Tx = f$ if and only if $K_{p,0} \neq \emptyset$ and if there exist $k_1 \in K_{p,x}$, $k_2 \in K_{p,-x}$ such that $k_1(-x) \geq -f$.

Proof: The condition is necessary, since every continuous, p-dominated operator $T : E \to F$ is a member of $\bigcap_{y \in E} K_{p,y}$. Conversely, choose concave, p-dominated mappings $k_o : U_o \to F$, $k_1 : x+U_1 \to F$, $k_2 : -x+U_2 \to F$ where U_o, U_1, U_2 are suitable convex zero-neighborhoods in E such that k_o, k_1, k_2 are continuous at 0, x and -x, respectively, and $k_1(x) \geq f$, $k_2(-x) \geq -f$. The operator $S : \mathbb{R}\cdot x \to F$, defined by $S(\lambda x) = \lambda f$, is p-dominated, since $f \leq k_1(x) \leq p(x)$ and $-f \leq k_2(-x) \leq p(-x)$. By Lemma 2.7 and Proposition 2.3 it suffices to show that the set $\{S(\lambda x) + p(y - \lambda x) : \lambda \in \mathbb{R}\}$ is bounded from below in C for every $y \in E$ and that $K_{q,0} \neq \emptyset$ for $q := p_S$. To this end, we select a symmetric, convex zero-neighborhood U in E and a real number $\varepsilon > 0$ such that $U + U \subset U_o$ and $[-\varepsilon x, \varepsilon x] \subset U$. If $y \in U$ and $\lambda \in [-\varepsilon, \varepsilon]$, the relation $y - \lambda x \in U + U \subset U_o$ yields

$$\begin{aligned} S(\lambda x) + p(y - \lambda x) &= \lambda f + p(y - \lambda x) \geq -\varepsilon|f| + k_o(y - \lambda x) \\ &= -\varepsilon|f| + k_o(\tfrac{1}{2}(2y) + \tfrac{1}{2}(-2\lambda x)) \\ &\geq -\varepsilon|f| + \tfrac{1}{2}k_o(2y) + \tfrac{1}{2}k_o(-2\lambda x) \end{aligned}$$

and

$$\begin{aligned} k_o(-2\lambda x) &= k_o(\tfrac{\varepsilon+\lambda}{2\varepsilon}(-2\varepsilon x) + \tfrac{\varepsilon-\lambda}{2\varepsilon}(2\varepsilon x)) \\ &\geq \tfrac{\varepsilon+\lambda}{2\varepsilon}k_o(-2\varepsilon x) + \tfrac{\varepsilon-\lambda}{2\varepsilon}k_o(2\varepsilon x) \\ &\geq \inf\{k_o(-2\varepsilon x),\ k_o(2\varepsilon x)\}. \end{aligned}$$

Consequently, we obtain $S(\lambda x) + p(y - \lambda x) \geq \frac{1}{2}k_o(2y) - a$, where $a := \varepsilon|f| - \frac{1}{2}\inf\{k_o(-2\varepsilon x),\ k_o(2\varepsilon x)\}$.

If $\lambda > \varepsilon$ and $y \in \varepsilon U_2$ the equality $\frac{1}{\lambda}y - x = \frac{\varepsilon}{\lambda}\cdot\frac{1}{\varepsilon}y - x \in U_2 - x$ implies

$$
\begin{aligned}
S(\lambda x) + p(y-\lambda x) &= \lambda f + \lambda p(\tfrac{1}{\lambda}y - x) \geq \lambda(f + k_2(\tfrac{1}{\lambda}y - x)) \\
&= \lambda(f + k_2(\tfrac{\varepsilon}{\lambda}(\tfrac{1}{\varepsilon}y - x) + (1-\tfrac{\varepsilon}{\lambda})(-x))) \\
&\geq \lambda(f + \tfrac{\varepsilon}{\lambda}k_2(\tfrac{1}{\varepsilon}y - x) + (1-\tfrac{\varepsilon}{\lambda})k_2(-x)) \\
&\geq \lambda f + \varepsilon k_2(\tfrac{1}{\varepsilon}y - x) + (\lambda - \varepsilon)(-f) \\
&= \varepsilon(k_2(\tfrac{1}{\varepsilon}y - x) + f).
\end{aligned}
$$

Finally, if $\lambda < -\varepsilon$ and $y \in \varepsilon U_1$ we conclude from the equality $\frac{1}{-\lambda}y + x = \frac{\varepsilon}{-\lambda} \cdot \frac{1}{\varepsilon}y + x \in U_1 + x$:

$$
\begin{aligned}
S(\lambda x) + p(y-\lambda x) &= -\lambda(-f + p(\tfrac{1}{-\lambda}y + x)) \geq -\lambda(-f + k_1(\tfrac{1}{-\lambda}y + x)) \\
&= -\lambda(-f + k_1(\tfrac{\varepsilon}{-\lambda}(\tfrac{1}{\varepsilon}y + x) + (1+\tfrac{\varepsilon}{\lambda})x)) \\
&\geq -\lambda(-f + \tfrac{\varepsilon}{-\lambda}k_1(\tfrac{1}{\varepsilon}y + x) + (1+\tfrac{\varepsilon}{\lambda})k_1(x)) \\
&\geq \lambda f + \varepsilon k_1(\tfrac{1}{\varepsilon}y + x) - (\lambda + \varepsilon)f \\
&= \varepsilon(k_1(\tfrac{1}{\varepsilon}y + x) - f).
\end{aligned}
$$

Furthermore, if $W := U \cap \varepsilon U_1 \cap \varepsilon U_2$, the mapping $k : W \to F$, given by

$$k(y) = \inf\{\tfrac{1}{2}k_o(2y) - a,\ \varepsilon(k_1(\tfrac{1}{\varepsilon}y + x) - f),\ \varepsilon(k_2(\tfrac{1}{\varepsilon}y - x) + f)\},$$

is concave and continuous at 0, since it is the greatest lower bound of finitely many concave mappings continuous at 0. Given $y \in E$ and $\rho \in \mathbb{R}_+^*$ such that $\rho y \in W$, we hence deduce from the inequality

$$S(\lambda x) + p(y - \lambda x) = \tfrac{1}{\rho}(S(\rho\lambda x) + p(\rho y - \rho\lambda x)) \geq \tfrac{1}{\rho}k(\rho x)$$

that the set $\{S(\lambda x) + p(y-\lambda x) : \lambda \in \mathbb{R}\}$ is bounded from below in C and that $p_S(y) = \inf_{\lambda \in \mathbb{R}}(S(\lambda x) + p(y - \lambda x)) \geq k(y)$ for all $y \in W$. ■

<u>2.9 Corollary</u>. (see [2], Thm. 2.11 and [32] théorème 5):
Let E be a locally convex Hausdorff space[+], H a finite-dimensional subspace of E and let $f : H \to \mathbb{R}$ be a p-dominated linear form on H, where $p : E \to \mathbb{R}_\infty$ is a sublinear functional. If p is l.s.c. at each

[+]) It is no loss of generality to assume the topology of E to be separated. (see the first lines of the proof of Thm. 2.11 in [2]).

point $h \in H$ and satisfies $p(h) > f(h)$ for all $h \in H$ such that $p(h) \neq -p(-h)$, then f can be extended to a continuous, p-dominated linear form on E.

<u>Proof</u>: Let S denote the set of all finite (pointwise) suprema $\sup\{f_1,\ldots,f_n\}$ of continuous, p-dominated linear forms $f_1,\ldots,f_n$ on E. We claim that $p(h) = \sup_{s\in S} s(h)$ for all $h \in H$.
Since this equality is evident whenever $p(h) = -p(-h)$, by Corollary 2.5,c, we may assume that $h \in H$ is such that $p(h) \neq -p(-h)$. From the inequality $p(h) + p(-h) \geq p(0) = 0$ it follows that $-p(-h) < p(h)$. Therefore, it suffices to show that $\sup_{s\in S} s(h) \geq \alpha$ for all $\alpha \in]-p(-h), p(h)[$.
To this end, choose $\varepsilon > 0$ such that $\alpha + \varepsilon < p(h)$ and $\alpha - \varepsilon > -p(-h)$. The functional p being l.s.c. at 0, h and -h, there exist convex neighborhoods U_o, U_1, U_2 of 0, h and -h, respectively, such that $p(x) \geq \alpha + \varepsilon$ for all $x \in U_1$, $p(x) \geq -\alpha + \varepsilon$ for all $x \in U_2$ and $p(x) > -1$ for all $x \in U_o$. Consequently, by Theorem 2.8, there is a continuous, p-dominated linear form $f_h : E \to \mathbb{R}$ attaining the value α at h. Since $f_h \in S$, we thus conclude $\sup_{s\in S} s(h) \geq \alpha$. If G is an algebraic complement of the linear space $L := \{h \in H : p(h) = -p(-h)\}$ in H and if S denotes the unit sphere of G with respect to some norm on G, then S is compact and the l.s.c. function $h \to p(h) - f(h)$ attains its minimum at some point $h_o \in S$. Since $h_o \in G$, it follows that $\delta := p(h_o) - f(h_o) > 0$. Therefore, $f + \frac{\delta}{2}$ is dominated by $p - \frac{\delta}{2}$ on S. By Dini's theorem there exists a sublinear functional s in the upward directed set S such that $s(h) \geq f(h)$ for all $h \in S$ and hence for all $h \in G$. The classical Hahn-Banach theorem then yields an s-dominated, linear extension $\tilde{f} : E \to F$ of $f|_G$. Since s is continuous, the continuity of $\tilde{f}$ follows. Finally, from the inequality $\tilde{f} \leq s \leq p$ we conclude $p(h) = -p(-h) \leq \tilde{f}(h) \leq p(h)$ for all $h \in L$, hence $p|_L = \tilde{f}|_L$, which proves $\tilde{f}$ to be a p-dominated extension of f. ∎

Given a locally convex vector space and a tight imbedding cone for a topological vector lattice F, let $p : E \to C$ be a sublinear mapping such that $K_{p,0} \neq \emptyset$. Then $K_{p,x} \neq \emptyset$ for each $x \in E$ by Proposition 2.4. Consequently, for each $x \in E$,

$$p^{\cap}(x) := \sup\{k(x) : k \in K_{p,x}\}$$

exists in C.

<u>2.10 Definition</u>: The mapping $p^{\cap} : E \to C$ is called <u>regularization</u> of p. If $p = p^{\cap}$, then p is said to be <u>regularized</u>.

In the next theorem we shall show that $p^{\cap}$ is the upper envelope of all p-dominated, continuous operators. For the proof we use the following

<u>2.11 Lemma</u>: Let p be a sublinear mapping from a locally convex vector space into a tight imbedding cone of some topological vector lattice F. Then the set $\{k(x) : k \in K_{p,x}\}$ is upward directed for each $x \in E$.

<u>Proof</u>: Consider two concave, p-dominated mappings $k_1 : x + U_1 \to F$, $k_2 : x + U_2 \to F$ continuous at x, where U_1, U_2 are convex zero-neighborhoods in E. F being Dedekind complete, the band projection P_1 from F onto the band generated by $(k_1(x) - k_2(x))^+$ is well-defined. Let P_2 denote the complementary band projection, i.e. $|P_1 f| \wedge |P_2 f| = 0$ and $P_1(f) + P_2(f) = f$ for all $f \in F$. Then we obtain

$$P_1((k_1(x) - k_2(x))^-) = 0 \quad \text{and} \quad P_2((k_1(x) - k_2(x))^+) = 0$$

consequently,

$$\begin{aligned} P_1(k_1(x)) + P_2(k_2(x)) &= P_1(k_1(x) + (k_1(x) - k_2(x))^-) + P_2((k_1(x) - k_2(x))^+ + k_2(x)) \\ &= P_1(k_1(x) \vee k_2(x)) + P_2(k_1(x) \vee k_2(x)) \\ &= k_1(x) \vee k_2(x). \end{aligned}$$

Hence, if we define $k : x + U_1 \cap U_2 \to F$ by

$$k(y) = P_1(k_1(y)) + P_2(k_2(y)),$$

then k is concave, continuous at x and $k(x) \geq k_1(x)$, $k(x) \geq k_2(x)$. Furthermore, for all $y \in x + U_1 \cap U_2$, we obtain

$$\begin{aligned} k(y) &= P_1(k_1(y)) + P_2(k_2(y)) \\ &\leq P_1(k_1(y) \vee k_2(y)) + P_2(k_1(y) \vee k_2(y)) \\ &= k_1(y) \vee k_2(y) \leq p(y). \end{aligned}$$

Therefore, $k \in K_{p,x}$. ∎

2.12 Theorem: Let p be a sublinear mapping from a locally convex vector space E into a tight imbedding cone C of a topological vector lattice F, and suppose that $K_{p,o} \neq \emptyset$. Then the regularization $p^{\cap}$ of p is the upper envelope of all p-dominated, continuous operators from E into F, i.e.,

$p^{\cap}(x) = \sup\{T(x) : T : E \to F$ p-dominated, linear, continuous$\}$

for all $x \in E$.

In particular, $p^{\cap}$ is sublinear.

Proof: Given $x \in E$ and $k \in K_{p,x}$, it suffices to show that there exists a continuous, p-dominated operator $T : E \to F$ such that $Tx \geq k(x)$. Note first that, by Proposition 2.3, there exists a p-dominated, continuous operator $T_o : E \to F$. The set $\{k'(x) : k' \in K_{p,x}\}$ being upward directed we can select a concave mapping $k' \in K_{p,x}$ such that $k'(x) \geq k(x) \vee T_o x =: f$, since $T_o \in K_{p,x}$. Finally, the inequality $-f \leq -T_o x = T_o(-x) \leq p(-x)$ shows that, by Theorem 2.8, there is a p-dominated continuous operator $T:E \to F$ satisfying $Tx = f \geq k(x)$. ∎

2.13 Application: Extension of positive operators

Let E be a normed vector space, F a Dedekind complete Banach lattice and let C be a tight imbedding cone of F possessing a maximal element ∞. For a linear subspace H of E and a continuous, positive operator $S : E \to F$, consider the mapping $p : E \to C$, defined by

$$p(e) = \sup_{\varepsilon>0} \inf_{\substack{h\in H \\ \|(e-h)^+\|\leq\varepsilon}} (Sh \vee Se) \quad \text{for each } e \in E,$$

where we use the convention $\inf \emptyset := \infty$.

The following theorem shows that the mapping p naturally occurs in the context of extensions of positive operators.

2.13.1 Theorem: The mapping p is sublinear, increasing and regularized. Moreover, $p(h) = Sh$ for all $h \in H$, S is p-dominated and every p-dominated operator $T : E \to F$ is a positive, linear extension of $S|_H$. Conversely, if a positive linear extension $T : E \to F$ of $S|_H$ satisfies the additional condition

$$\text{(2.13.2)} \qquad \inf_{\varepsilon>0} \sup_{\substack{e\in E \\ \|e\|\leq\varepsilon}} Te = 0 \quad \text{in } C,$$

then T is p-dominated.

Proof: The mapping p is clearly positively homogeneous. In order to prove the subadditivity, let $e_1, e_2 \in E$, $\varepsilon > 0$ and $h_1, h_2 \in H$ be such that $\|(e_1 - h_1)^+\| \leq \varepsilon$, $\|(e_2 - h_2)^+\| \leq \varepsilon$. Then we have $\|(e_1+e_2-(h_1+h_2))^+\| \leq 2\varepsilon$, hence

$$\inf_{\substack{h\in H \\ \|(e_1+e_2-h)^+\|\leq 2\varepsilon}} (Sh \vee S(e_1 + e_2)) \leq Sh_1 \vee Se_1 + Sh_2 \vee Se_2.$$

Passing to the greatest lower bounds on the right side, we conclude

$$\inf_{\substack{h\in H \\ \|(e_1+e_2-h)^+\|\leq 2\varepsilon}} (Sh \vee S(e_1+e_2)) \leq \inf_{\substack{h_1\in H \\ \|(e_1-h_1)^+\|\leq\varepsilon}} (Sh_1 \vee Se_1) + \inf_{\substack{h_2\in H \\ \|(e_2-h_2)^+\|\leq\varepsilon}} (Sh_2 \vee Se_2)$$

$$\leq p(e_1) + p(e_2).$$

Since $\varepsilon > 0$ was arbitrary, it follows that $p(e_1 + e_2) \leq p(e_1) + p(e_2)$.
From the inclusion $\{h \in H : \|(e-h)^+\| \leq \varepsilon\} \supset \{h \in H : \|(e'-h)^+\| \leq \varepsilon\}$ valid for each $e, e' \in E$ such that $e \leq e'$ and all $\varepsilon > 0$, we immediately deduce that p is increasing.
To show the equality $p^\cap = p$, fix $e \in E$ and let $\varepsilon > 0$ be given. For each $e' \in E$ such that $\|(e-e')^+\| \leq \frac{\varepsilon}{2}$ and each $h \in H$ satisfying $\|(e'-h)^+\| \leq \frac{\varepsilon}{2}$ we deduce from the inequalities $\|(e-h)^+\| \leq \|(e-e')^+\| + \|(e'-h)^+\| \leq \varepsilon$ and $Sh \vee Se - S(e-e')^+ \leq Sh \vee Se'$ that

$$\inf_{\substack{h \in H \\ \|(e-h)^+\| \leq \varepsilon}} (Sh \vee Se) - S(e-e')^+ \leq \inf_{\substack{h \in H \\ \|(e'-h)^+\| \leq \frac{\varepsilon}{2}}} (Sh \vee Se') \leq p(e').$$

Hence, if we set

$$k_\varepsilon(e') = \inf_{\substack{h \in H \\ \|(e-h)^+\| \leq \varepsilon}} (Sh \vee Se) - S(e-e')^+$$

for each $e' \in \{x \in E : \|e - x\| \leq \frac{\varepsilon}{2}\}$, then $k_\varepsilon \in \boldsymbol{K}_{p,e}$ for all $\varepsilon > 0$. From the equality $\sup\{k_\varepsilon(e) : \varepsilon > 0\} = p(e)$ it therefore follows that $p^\cap(e) = p(e)$.
By the definition of p we have $Se \leq p(e)$ for all $e \in E$. Moreover, if $h \in H$, then

$$\inf_{\substack{h' \in H \\ \|(h-h')^+\| \leq \varepsilon}} (Sh' \vee Sh) \leq Sh \vee Sh = Sh \quad \text{for all } \varepsilon > 0,$$

Consequently, $p(h) \leq Sh$, which yields $p(h) = Sh$.
Finally, let $T : E \to F$ be a p-dominated operator. Then $Th \leq p(h) = Sh$ and $T(-h) \leq p(-h) = S(-h)$, hence $Th = Sh$ for each $h \in H$. From the inequality

$$T(-e) \leq p(-e) \leq p(0) = 0,$$

valid for every $e \in E_+$, we thus conclude that T is a positive linear extension of $S|_H$.
Conversely, if $T : E \to F$ is a positive linear extension of $S|_H$ satisfying condition 2.13.2, then we obtain

$$Te \leq Th + T((e-h)^+) \leq Sh + \sup_{\substack{h' \in H \\ \|(e-h')^+\| \leq \varepsilon}} T((e-h')^+)$$

for $e \in E$, $\varepsilon > 0$ and $h \in H$ such that $\|(e-h)^+\| \leq \varepsilon$. Hence, if we set

$$f_\varepsilon := \sup_{\substack{h' \in H \\ \|(e-h')^+\| \leq \varepsilon}} T((e-h')^+),$$

the inequality $Te \leq \inf_{\substack{h \in H \\ \|(e-h)^+\| \leq \varepsilon}} (Sh \vee Se) + f_\varepsilon$ yields $Te \leq p(e) + f_\varepsilon$.

From Lemma 1.3,iv we thus conclude

$$Te \leq \inf_{\varepsilon > 0} (p(e) + f_\varepsilon) = p(e) + \inf_{\varepsilon > 0} f_\varepsilon = p(e),$$

since $\inf_{\varepsilon > 0} f_\varepsilon = 0$ by 2.13.2. ∎

2.13.3 Consequences

i) Given a compact topological space X, let $E = C(X)$ and choose a tight imbedding cone C of a topological vector lattice F such that C has a maximal element. Then every positive operator $T : E \to F$ satisfies condition 2.13.2. Indeed, if 1 denotes the constant function with value 1 on X, we have

$$\inf_{\varepsilon > 0} \sup_{\|e\| \leq \varepsilon} Te = \inf_{\varepsilon > 0} \varepsilon \cdot T(1) = 0.$$

Consider a positive operator $S : E \to F$ and a linear subspace H of E. For each $e \in E$, the relation

$$\inf_{\substack{h \in H \\ \|(e-h)^+\| \leq \varepsilon}} Sh = \inf_{\substack{h \in H \\ h \geq e - \varepsilon \cdot 1}} Sh \geq Se - \varepsilon S1$$

yields

$$\sup_{\varepsilon > 0} \inf_{\substack{h \in H \\ \|(e-h)^+\| \leq \varepsilon}} Sh \geq Se.$$

Therefore, the sublinear mapping p introduced in 2.13.1 can be simplified to

$$p(e) = \sup_{\varepsilon>0} \left(\left(\inf_{\substack{h\in H \\ \|(e-h)^+\|\leq\varepsilon}} Sh \right) \vee Se \right) = \left(\sup_{\varepsilon>0} \inf_{\substack{h\in H \\ \|(e-h)^+\|\leq\varepsilon}} Sh \right) \vee Se = \sup_{\varepsilon>0} \inf_{\substack{h\in H \\ h\geq e-\varepsilon 1}} Sh.$$

An application of 2.12 and 2.13.1 hence shows that

$$p(e) = \sup\{Te : T : E \to F \text{ pos. lin. extension of } S|_H\}$$

for each $e \in E$.

Consequently, the set

$$M := \{e \in E : p(e) = Se \text{ and } p(-e) = S(-e)\}$$

is the linear subspace of E, on which all positive extensions of $S|_H$ coincide. Since S is p-dominated, it follows that $e \in M$ if and only if $p(e) \leq Se$ and $p(-e) \leq S(-e)$, or, equivalently, if

$$\inf_{\substack{h\in H \\ h\geq e-\varepsilon 1}} Sh \leq Se \quad \text{and}$$

$$-\sup_{\substack{h'\in H \\ h'\leq e+\varepsilon 1}} Sh' = \inf_{\substack{h'\in H \\ -h'\geq -e-\varepsilon 1}} S(-h') = \inf_{\substack{h\in H \\ h\geq -e-\varepsilon 1}} Sh \leq S(-e) = -Se.$$

Therefore, the condition

$$\inf_{\substack{h\in H \\ h\geq e-\varepsilon 1}} Sh \leq Se \leq \sup_{\substack{h\in H \\ h\leq e+\varepsilon 1}} Sh \quad \text{for all } \varepsilon > 0$$

is necessary and sufficient for the equality $Te = Se$ to hold for every positive linear extension T of $S|_H$.

ii) Given a σ-finite measure space $(\Omega, \mathcal{A}, \mu)$, let $C := C_\infty$ denote the sup-completion of $F := L^\infty(\mu)$ (see Example 1.2,d). If E is an arbitrary normed vector lattice, then condition (2.13.2) is again satisfied for each positive, continuous operator $T : E \to F$, since

$$\inf_{\varepsilon>0} \sup_{\substack{e\in E \\ \|e\|\leq\varepsilon}} Te = \inf_{\varepsilon>0} \|T\| \cdot \varepsilon \cdot 1 = 0.$$

Using the same arguments as in (i) we obtain

$$\{e \in E : Te = Se \text{ for each pos. extension } T \text{ of } S|_H\}$$
$$= \{e \in E : \inf_{\substack{h\in H \\ \|(e-h)^+\|\le\varepsilon}} Sh \le Se \le \sup_{\substack{h\in H \\ \|(e-h)^-\|\le\varepsilon}} Sh \text{ for all } \varepsilon > 0\}.$$

iii) Given a normed vector lattice E and a real number $q \in [1,\infty[$, let $F = \ell^q$ be the space of all real sequences $x = (\xi_n)_{n\in\mathbb{N}}$ satisfying

$$\|x\| := (\sum_n |\xi_n|^q)^{1/q} < \infty.$$

The sup-completion C of ℓ^q can be identified with the tight imbedding cone of all $\mathbb{R}_\infty$-valued sequences possessing an ℓ^q-minorant. For each continuous positive operator $T : E \to F$ the equality

$$\inf_{\varepsilon>0} \sup_{\substack{e\in E \\ \|e\|\le\varepsilon}} Te = \inf_{\varepsilon>0} \|T\| \cdot x_\varepsilon = 0$$

holds in C, where $x_\varepsilon \in C$ denotes the constant sequence with value ε. Thus, condition (2.13.2) is again satisfied. Using the same notations as in Example (i) $\inf_{\substack{h\in H \\ \|(e-h)^+\|\le\varepsilon}} Sh$ does no longer exist <u>in C</u> for all $\varepsilon > 0$, but forming pointwise infima we obtain

$$\inf_{\substack{h\in H \\ \|(e-h)^+\|\le\varepsilon}} Sh(n) \ge Se(n) - \varepsilon \quad \text{for each } n \in \mathbb{N},\ e \in E,$$

where Sh(n) and Se(n) denotes the n-th member of the respective sequences. Consequently,

$$p(e)(n) = \sup_{\varepsilon>0} \inf_{\substack{h\in H \\ \|(e-h)^+\|\le\varepsilon}} (Sh \vee Se)(n) = \sup_{\varepsilon>0} \inf_{\substack{h\in H \\ \|(e-h)^+\|\le\varepsilon}} Sh(n)$$

for all $e \in E$, $n \in \mathbb{N}$. A similar argument as in (i) therefore yields

$$\{e \in E : Te = Se \text{ for all pos. extensions } T \text{ of } S|_H\}$$
$$= \{e \in E : \inf_{\substack{h\in H \\ \|(e-h)^+\|\le\varepsilon}} Sh(n) \le Se(n) \le \sup_{\substack{h\in H \\ \|(e-h)^-\|\le\varepsilon}} Sh(n) \text{ for all } \varepsilon > 0, n \in \mathbb{N}\}.$$

2.14 Remark. The extension theorems of this chapter only provide a first approach to a satisfactory solution of extension problems. Thus, e.g., it is often difficult to construct vector-valued concave mappings. Moreover, many extension problems for operators can not be formulated by means of vector-valued sublinear mappings. Consider, for example, the Dedekind complete Banach lattice $L^1(\lambda)$, where λ denotes the Lebesgue measure on $[0,1]$. The sup-completion of $L^1(\lambda)$ is then isomorphic to the tight imbedding cone C_1 described in 1.2,d.

Suppose, there were a sublinear mapping $p : L^1(\lambda) \to C_1$ satisfying the following condition:
For a continuous operator $T : L^1(\lambda) \to L^1(\lambda)$ to be p-dominated, it is necessary and sufficient that T is contractive, i.e. $\|T\| \leq 1$.
As a consequence, the inequality

$$\sup\{Tf : T : L^1(\lambda) \to L^1(\lambda), \|T\| \leq 1\} \leq p(f)$$

holds for each $f \in L^1(\lambda)$. We shall show that this is impossible.

Let $g_{n,k} \in L^1(\lambda)$ denote the equivalence class of the characteristic function of the interval $[\frac{k-1}{n}, \frac{k}{n}]$ for each $k,n \in \mathbb{N}$ such that $k \leq n$. If $f \in L^1(\lambda)$, $f \neq 0$, the operator $T_{k,n} : L^1(\lambda) \to L^1(\lambda)$ defined by

$$T_{k,n}(g) = n \cdot g_{k,n} \int g \cdot \operatorname{sgn}(f) d\lambda,$$

where $\operatorname{sgn}(f) \in L^\infty(\lambda)$ is the (equivalence class of the) signum function of f, is continuous, contractive and

$$T_{k,n}(f) = n \cdot g_{k,n} \|f\|.$$

Furthermore, if $\mathbb{1}$ denotes the equivalence class of the constant function with value 1, then $\sup\{T_{k,n}(f) : 1 \leq k \leq n\} = n \cdot \|f\| \cdot \mathbb{1}$. Therefore, from the equality

$$\sup\{T_{k,n}(f) : 1 \leq k \leq n,\ n \in \mathbb{N}\} = \infty \cdot \mathbb{1} \in C_1,$$

we deduce that $p(f) = \infty \cdot \mathbb{1}$. Since $f \in L^1(\lambda) \setminus \{0\}$ was chosen arbitrarily, we thus conclude that <u>every</u> continuous operator $T : L^1(\lambda) \to L^1(\lambda)$ is p-dominated contradicting the assumption that T should be contractive.

By the way, things cannot be improved by choosing other tight imbedding cones for $L^1(\lambda)$, since each tight imbedding cone is lattice cone isomorphic to a lattice subcone of the sup-completion C_1.

We thus need some additional concepts for regularizing sublinear mappings and a new approach to the solution of extension problems for operators including norm preserving extensions. The duality theory of tensor products introduced in the next chapter will prove efficient on this way.

<u>Final remarks to Chapter 2</u>

a) Lemma 2.2 and Proposition 2.3 are closely related to so-called sandwich theorems for convex and concave mappings (see [56], [79]). As the statements are only auxiliary results for Theorem 2.8 and since the formulation in the context of imbedding cones seems to be new, we preferred to supply complete proofs.
b) Conversely, it is almost obvious how to derive sandwich theorems from Theorem 2.8. Preferring a quick development of the main results, we omit the details.

3. Bisublinear and subbilinear functionals

Throughout this chapter, let E and F denote (real) vector spaces.

3.1 Definiton: A function $p : E \times F \to \mathbb{R}_\infty$ which is sublinear in each variable separately, is called bisublinear. Given a bisublinear function p we define $p^\otimes : E \otimes F \to \mathbb{R} \cup \{-\infty,\infty\}$ by the equality

$$p^\otimes(t) = \inf\{ \sum_{i\in I} p(e_i,f_i) : (e_i,f_i)_{i\in I} \text{ finite family in } E \times F \text{ such that } t = \sum_i e_i \otimes f_i\}.$$

If $p^\otimes(t) > -\infty$ for all $t \in E \otimes F$, then $p^\otimes$ is obviously sublinear.

3.2 Example: For a bilinear form $b : E \times F \to \mathbb{R}$, $b^\otimes$ is the uniquely determined linear form on $E \otimes F$ satisfying the equality $b^\otimes(e \otimes f) = b(e,f)$ for each $(e,f) \in E \times F$.

3.3 Lemma: Given a bisublinear functional $p : E \times F \to \mathbb{R}_\infty$ and a bilinear form $b : E \times F \to \mathbb{R}$, the following are equivalent

i) $b \leq p$,

ii) $b^\otimes \leq p^\otimes$,

iii) $b^\otimes(e \otimes f) \leq p^\otimes(e \otimes f)$ for all $(e,f) \in E \times F$.

The proof is clear.

3.4 Proposition: Let F_+ be a generating cone in F, i.e. $F_+ - F_+ = F$ and let $p : E \times F \to \mathbb{R}_\infty$ be sublinear. If $p(E \times F_+) \subset \mathbb{R}$, then $p^\otimes(E \otimes F) \subset \mathbb{R} \cup \{-\infty\}$ and the equality

$$p^\otimes(t) = \sup\{b^\otimes(t) : b \text{ p-dominated bilinear form on } E \times F\}$$

holds for each $t \in E \otimes F$, where we use the convention $\sup \emptyset := -\infty$.

Proof: Since the function p attains only finite values on $E \times F_+$, it follows from the relation $F = F_+ - F_+$ that $p^{\otimes}(t) < +\infty$ for all $t \in E \otimes F$. Suppose first that $p^{\otimes}(t) = -\infty$ for some $t \in E \otimes F$. Then the inequality

$$p^{\otimes}(t') \leq p^{\otimes}(t) + p^{\otimes}(t'-t) = -\infty$$

implies $p^{\otimes}(t') = -\infty$ for each $t' \in E \otimes F$. On the other hand, by Lemma 3.3, p-dominated bilinear forms on $E \times F$ cannot exist in this case. If $p^{\otimes}(t) > -\infty$ for all $t \in E \otimes F$, the $p^{\otimes}$ is a real-valued sublinear form. Hence the assertion follows from the Hahn-Banach theorem. ∎

3.5 Corollary: With the same assumptions as in Proposition 3.4 $p^{\otimes}$ is sublinear if and only if there exists a bilinear minorant b of p.

3.6 Definition: If F_+ is a subcone of F, a bisublinear functional $p : E \times F \to \mathbb{R}_\infty$ is called subbilinear on $E \times F_+$ provided that

$$p(e,f) = p^{\otimes}(e \otimes f) \quad \text{for all} \quad (e,f) \in E \times F_+.$$

A subbilinear form is a functional which is subbilinear on $E \times F$.

3.7 Examples

i) Let $r : E \to \mathbb{R}$ and $s : F \to \mathbb{R}$ be semi-norms. Then

$$p(e,f) = r(e) \cdot s(f) \qquad (e \in E,\ f \in F)$$

is a subbilinear form on $E \times F$. Using Proposition 3.4 with $F_+ := F$ it suffices to show that for each $(e,f) \in E \times F$ there is a p-dominated bilinear form $b : E \times F \to \mathbb{R}$ satisfying $b(e,f) = r(e) \cdot s(f)$. To this end, choose linear forms $\ell_1 : E \to \mathbb{R}$, $\ell_2 : F \to \mathbb{R}$ such that $\ell_1 \leq r$, $\ell_2 \leq s$ and $\ell_1(e) = r(e)$, $\ell_2(f) = s(f)$, which is possible by the Hahn-Banach theorem. Hence the bilinear form $b : E \times F \to \mathbb{R}$, defined by $b(e',f') = \ell_1(e')\ell_2(f')$ for all $e' \in E$, $f' \in F$, is p-dominated and $b(e,f) = r(e) \cdot s(f) = p(e,f)$.

ii) Every bilinear form $b : E \times F \to \mathbb{R}$ is a subbilinear form.

iii) If E, F are <u>normed vector lattices</u>, then the bisublinear functional $(e,f) \to \|e^+\| \cdot \|f^+\|$ is <u>not</u> a subbilinear form. In fact, every subbilinear form $p : E \times F \to \mathbb{R}_\infty$ satisfies the equality

$$p(e,f) = p^{\otimes}(e \otimes f) = p^{\otimes}((-e) \otimes (-f)) = p(-e,-f)$$

for all $(e,f) \in E \times F$.

Consider an operator T from a vector space E into the algebraic dual F^* of a vector space F. We define the canonically associated bilinear form $b_T : E \times F \to \mathbb{R}$ by the equality

$$b_T(e,f) = Te(f) \quad (e \in E,\ f \in F).$$

The mapping $T \to b_T$ is an isomorphism of the vector space of all linear mappings of E into F^* onto the vector space of all bilinear forms on $E \times F$. For simplicity, we shall write $T^{\otimes}$ rather than $b_T^{\otimes}$ for an operator $T : E \to F^*$. Moreover, by abuse of language, we shall use the phrase "<u>T is p-dominated</u>" instead of "$T^{\otimes}$ is $p^{\otimes}$-dominated" for a bisublinear functional $p : E \times F \to \mathbb{R}_\infty$.

<u>3.8 Theorem</u>: Let Φ be a sublinear mapping from a vector space E into a Dedekind complete vector lattice G. If F is a linear subspace of the algebraic dual G^* of G and if $p : E \times F \to \mathbb{R}_\infty$ is given by

$$p(e,f) = \begin{cases} f(\Phi(e)) & \text{for } e \in E,\ f \in F_+ \\ 0 & \text{for } e = 0,\ f \in F \setminus F_+ \\ \infty & \text{for } e \in E \setminus \{0\},\ f \in F \setminus F_+, \end{cases}$$

then p is subbilinear on $E \times F_+$. If, in addition, G is a Banach lattice and F a dense vector sublattice of the topological dual G' of G (endowed with the norm topology), then $T(E) \subset F' = G''$ for every p-dominated operator $T : E \to F^*$, where continuous linear forms on G' and on the dense subspace F are identified.

Proof: Given $(e,f) \in E \times F_+$ let $T : E \to G$ be a Φ-dominated operator such $Te = \Phi(e)$. Then T is obviously p-dominated and $f(Te) = p(e,f)$. Using Proposition 3.4 we obtain $p(e,f) = p^{\otimes}(e \otimes f)$. Hence p is subbilinear on $E \times F_+$. If G is a Banach lattice and F a dense vector sublattice of G', it follows that

$$\begin{aligned} Te(f) = Te(f^+) + T(-e)(f^-) &\leq f^+(\Phi(e)) + f^-(\Phi(-e)) \\ &\leq \|f\| (\|\Phi(e)\| + \|\Phi(-e)\|) \end{aligned}$$

for every p-dominated operator $T : E \to F^*$ and all $e \in E$, $f \in F$. Hence $Te \in F' = G''$. ∎

3.9 Corollary: Let Φ be a sublinear mapping from a vector space E into a Banach lattice G with order continuous norm. For a dense vector sublattice F of G' define the bisublinear functional $p : E \times F \to \mathbb{R}_\infty$ as in Theorem 3.8. Then, if $J : G \to G''$ denotes the natural imbedding, the relations

$$T(E) \subset J(G) \quad \text{and} \quad J^{-1} \circ T \leq \Phi$$

hold for every p-dominated operator $T : E \to F^*$.

Proof: Since a Banach lattice with order continuous norm is Dedekind complete (see [66], Ch. 2, Th. 5.10, Cor. 1), we deduce $T(E) \subset G''$ from Theorem 3.8. Furthermore, if $e \in E$ is arbitrary, then from the inequality $Te(f) \leq p(e,f) = f(\Phi(e))$ valid for all $f \in F_+$ we conclude

$$\begin{aligned} Te(g') &\leq g'(\Phi(e)) \quad \text{and} \\ T(-e)(g') &\leq g'(\Phi(-e)) \quad \text{for all } g' \in G'_+ , \end{aligned}$$

using the denseness of F_+ in G'_+. Consequently,

$$g'(-\Phi(-e)) \leq Te(g') \leq g'(\Phi(e)) \quad \text{for all } g' \in G'_+ ,$$

which shows that Te lies in the vector lattice ideal generated by $J(G)$. Since G has order continuous norm, we thus obtain $Te \in J(G)$ (see [66], Ch. 2, Th. 5.10). The rest of the assertion is evident. ∎

3.10 Examples and Remarks

i) Given two vector spaces E, F and a generating cone F_+ in F let $p : E \times F \to \mathbb{R}_\infty$ be a bisublinear functional, which is real-valued and subbilinear on $E \times F_+$. For $f \in F_+$ consider a bilinear form b on $E \times \mathbb{R}f$ such that

$$b(e,f) \leq p(e,f) \quad \text{for all } e \in E.$$

Then there exists a p-dominated bilinear extension b_o of b on $E \times F$. Indeed, $p^\otimes$ is real-valued and sublinear on $E \otimes F$ and

$$b^\otimes(\sum_{i \in I} e_i \otimes \lambda_i f) = b(\sum_i \lambda_i e_i, f) \leq p(\sum_i \lambda_i e_i, f)$$
$$= p^\otimes((\sum_i \lambda_i e_i) \otimes f) = p^\otimes(\sum_i e_i \otimes \lambda_i f)$$

for each finite family $(e_i, \lambda_i)_{i \in I}$ in $E \times \mathbb{R}$. Thus, the Hahn-Banach theorem and Lemma 3.3 complete the proof.

ii) The reader should keep in mind, however, that the conclusion of (i) is <u>no</u> <u>longer</u> <u>true</u> when the domain $E \times \mathbb{R}f$ of b is replaced by $E_1 \times F_1$ for arbitrary linear subspaces E_1 of E and F_1 of F. In fact, the inequality $b^\otimes(t) \leq p^\otimes(t)$ is not valid for <u>all</u> $t \in E_1 \otimes F_1$, in general, even if p is a subbilinear form.
Looking for a counterexample, consider two normed vector spaces E, G, and set

$$p(e,f) := \|e\| \, \|f\| \quad \text{for all } (e,f) \in E \times G'.$$

Then p is a subbilinear form by Example 3.7,i. Let T be an operator from a linear subspace H of E into the algebraic dual F^* of $F := G'$. Then T is p-dominated iff $T(H) \subset F' = G''$ and T is contractive, i.e. $\|T\| \leq 1$. But, even for finite-dimensional normed spaces E and G, contractive, linear extensions of a contractive operator $T : H \to G$ do not exist, in general.

3.11 Application (see [27]): Let Φ be a sublinear mapping from a vector space E into a Banach lattice G with order continuous norm. If f is a positive linear form on G and φ a linear form on E such that $\varphi \leq f \circ \Phi$, then there exists a Φ-dominated linear operator $T : E \to G$ satisfying $\varphi = f \circ T$.

Proof: For F := G', consider the bisublinear functional $p : E \times F \to \mathbb{R}_\infty$ defined in the same way as in Theorem 3.8. The bilinear form $b : E \times \mathbb{R}f \to \mathbb{R}$ given by

$$b(e, \lambda f) = \lambda \varphi(e) \qquad (e \in E,\ \lambda \in \mathbb{R})$$

is obviously p-dominated. Hence the assertion follows from Corollary 3.9 and Example 3.10,i. ∎

As a consequence of 3.11 we immediately deduce the following version of Strassen's theorem (see [33], [42], [76]):

3.12 Theorem (Strassen): Given a probability space $(\Omega, \mathcal{A}, \mu)$ and a sublinear mapping Φ from a vector space E into $\mathcal{L}^\infty(\mu)$ (notation of 1.2,d), let $\bar{\Phi}(e)$ denote the equivalence class of $\Phi(e)$ in $L^\infty(\mu)$ for each $e \in E$. Suppose that there exists a positive linear lifting $\varphi : L^\infty(\mu) \to \mathcal{L}^\infty(\mu)$ and a μ-negligible set $N \in \mathcal{A}$ such that $\varphi(\bar{\Phi})(e)(\omega) \leq \Phi(e)(\omega)$ for all $\omega \in \Omega \setminus N$ and $e \in E$. If ℓ is a linear form on E satisfying $\ell(e) \leq \int \Phi(e)\,d\mu$ for all $e \in E$, then there is an operator $T : E \to \mathcal{L}^\infty(\mu)$ such that

$$\ell(e) = \int Te\,d\mu \quad \text{and} \quad Te(\omega) \leq \Phi(e)(\omega)$$

for each $e \in E$, $\omega \in \Omega \setminus N$.

Proof: Since $L^\infty(\mu) \subset L^1(\mu)$, $\bar{\Phi}$ defines a sublinear mapping from E into $L^1(\mu)$. Consider the linear form $f \in L^1(\mu)'$ given by

$$f(g) = \int g\,d\mu \qquad (g \in L^1(\mu)).$$

From the inequality $\ell \leq f \circ \bar{\Phi}$ we conclude that there exists a $\bar{\Phi}$-dominated operator $T_1 : E \to L^1(\mu)$ such that $\ell = f \circ T_1$ using 3.11. The inclusion $\bar{\Phi}(E) \subset L^\infty(\mu)$ therefore yields $T_1(E) \subset L^\infty(\mu)$. Thus, the operator $T := \varphi \circ T_1$ has the demanded properties. ∎

Note that, although a bisublinear form $p : E \times F \to \mathbb{R}$ may satisfy the inequality $\sum_i p(e_i, f_i) \geq 0$ for every finite family $(e_i, f_i)_{i \in I}$ in $E \times F$ such that $\sum_i e_i \otimes f_i = 0$, p need not be a subbilinear form. Consequently, the conclusion of 3.10,i does not hold for bisublinear forms of this type. As a counterexample, consider the bisublinear form

$$p(e,f) := \|e^+\| \; \|f\| \qquad (e \in E, \; f \in F),$$

where E is a normed vector lattice and F is a normed space. Then the only p-dominated operator $T : E \to F^*$ is the constant mapping $T = 0$. Indeed, if T is p-dominated, we obtain

$$Te(f) \leq \|e^+\| \; \|f\| = 0 \qquad \text{for all } e \in -E_+, \; f \in F.$$

Te being a linear form, we deduce $Te = 0$ for all $e \in -E_+$, which yields $T = 0$ since $E = E_+ - E_+$.

Furthermore, we can choose an element $f \in F$ satisfying $\|f\| = 1$ and a linear form ℓ on E, $\ell \not\equiv 0$, such that $\ell(e) \leq \|e^+\|$ for each $e \in E$. From the inequality

$$\ell(e) \leq \|e^+\| = p(e,f) \qquad \text{valid for all } e \in E,$$

we conclude that the bilinear form $b : E \times \mathbb{R}f \to \mathbb{R}$ given by

$$b(e, \lambda f) = \lambda \cdot \ell(e) \qquad (e \in E, \; \lambda \in \mathbb{R})$$

is p-dominated. There is, however, no p-dominated bilinear extension $b_o : E \times F \to \mathbb{R}$ of b, else the associated operator $T : E \to F^*$ satisfying

$$Te(f') = b_o(e, f') \qquad (e \in E, \; f' \in F)$$

had to be the constant mapping $T = 0$ contradicting the assumptions $\ell \not\equiv 0$, $\|f\| = 1$.

We are now prepared to resume the discussion of the open problems mentioned at the end of Chapter 2. In particular, we wish to obtain a better description of regularizations. As a first step we shall combine the tensor product approach and the vector-valued extension theory of Chapter 2.:

<u>3.13 Lemma</u>: Given a normed vector space F, let C denote the set of all l.s.c., additive, positively homogeneous functions $\varphi : F_+ \to \mathbb{R}_\infty$. Endowed with pointwise operations and order, C is a tight imbedding cone of the topological dual F' of F isomorphic to sup-completion of F'. Moreover,

$$\varphi(f) = \sup\{\ell(f) : \ell \in F', \ell|_{F_+} \ \varphi\text{-dominated}\}$$

for each $f \in F$.

<u>Proof</u>: Obviously, C is an ordered cone. To each functional $\varphi \in C$ we associate the l.s.c., sublinear functional $\tilde{\varphi} : F \to \mathbb{R}_\infty$, defined by

$$\tilde{\varphi}(f) = \begin{cases} \varphi(f), & \text{if } f \in F_+ \\ \infty & \text{else.} \end{cases}$$

By Theorem 2.12, we obtain

$$\begin{aligned} \varphi(f) = \tilde{\varphi}(f) &= \sup\{\ell(f) : \ell \in F', \ell \ \tilde{\varphi}\text{-dominated}\} \\ &= \sup\{\ell(f) : \ell \in F', \ell|_{F_+} \ \varphi\text{-dominated}\} \end{aligned}$$

for each $f \in F_+$. Note that $L_\varphi := \{\ell \in F' : \ell|_{F_+} \ \varphi\text{-dominated}\}$ is upward directed. Indeed, for $\ell_1, \ell_2 \in L_\varphi$ and each $f \in F_+$ we have

$$\begin{aligned} (\ell_1 \vee \ell_2)(f) &= \sup\{\ell_1(f_1) + \ell_2(f_2) : f_1, f_2 \in F_+ \text{ such that } f_1 + f_2 = f\} \\ &\leq \sup\{\varphi(f_1) + \varphi(f_2) : f_1, f_2 \in F_+ \text{ such that } f_1 + f_2 = f\} \\ &= \varphi(f), \end{aligned}$$

which implies that $\ell_1 \vee \ell_2 \in L_\varphi$. Consequently, the sets

$$L_{\varphi_1} \vee L_{\varphi_2} = \{\ell_1 \vee \ell_2 : \ell_1 \in L_{\varphi_1}, \ \ell_2 \in L_{\varphi_2}\} \text{ and}$$

$$L_{\varphi_1} \wedge L_{\varphi_2} = \{\ell_1 \wedge \ell_2 : \ell_1 \in L_{\varphi_1}, \ell_2 \in L_{\varphi_2}\}$$

are upward directed for each $\varphi_1, \varphi_2 \in C$. From this fact we immediately deduce that $\sup(L_{\varphi_1} \vee L_{\varphi_2})$ is the least upper bound and $\sup(L_{\varphi_1} \wedge L_{\varphi_2})$ is the greatest lower bound of φ_1 and φ_2. In particular, C is a lattice cone.

Consider a real-valued, additive, positively homogeneous functional $\rho : F_+ \to \mathbb{R}$. If ℓ_ρ is the uniquely determined linear extension of ρ to F, then the equality

$$\ell_\rho(f) = \rho(f^+) - \rho(f^-) \quad \text{for all } f \in F,$$

shows that ℓ_ρ is continuous if ρ is continuous at 0. In this case, however, ρ is continuous at each point of F_+.

Therefore, the mapping Λ from F' into the vector lattice C_o of all additively invertible elements in C, defined by

$$\Lambda(\ell) := \ell|_{F_+} \quad (\ell \in F'),$$

is onto and hence a vector lattice isomorphism.

In order to show that C is tight, let φ be a given l.s.c. functional in C dominated by some $\varphi_1 \in C_o$. Since $\varphi(0) = \varphi_1(0) = 0$, φ is u.s.c. and therefore continuous at 0. It follows that φ is continuous or, equivalently, $\varphi \in C_o$.

Thus, C will be a tight imbedding cone of F' provided that the equality $\varphi_o + \varphi \wedge \ell = (\varphi_o + \varphi) \wedge (\varphi_o + \ell)$ holds for each $\varphi_o, \varphi \in C$, $\ell \in F'$. To prove this, note first that the functional $\rho_{\varphi_1,\varphi_2} : F_+ \to \mathbb{R}_\infty$, $\varphi_1, \varphi_2 \in C$, defined by

$$\rho_{\varphi_1,\varphi_2}(f) = \inf\{\varphi_1(g) + \varphi_2(f - g) : g \in F_+, g \leq f\}$$

is positively homogeneous. Moreover, $\rho_{\varphi_1,\varphi_2}$ is additive, since

$$\rho_{\varphi_1,\varphi_2}(f_1 + f_2) = \inf\{\varphi_1(g) + \varphi_2(f_1 + f_2 - g) : g \in F_+, g \leq f_1 + f_2\}$$

$$= \inf\{\varphi_1(g_1+g_2)+\varphi_2(f_1+f_2-g_1+g_2) : g_1, g_2 \in F_+, g_1 \leq f_1, g_2 \leq f_2\}$$

$$= \inf\{\varphi_1(g_1) + \varphi_2(f_1 - g_1) + \varphi_1(g_2) + \varphi_2(f_2 - g_2) : g_1, g_2 \in F_+, \\ g_1 \leq f_1, g_2 \leq f_2\}$$

$$= \rho_{\varphi_1,\varphi_2}(f_1) + \rho_{\varphi_1,\varphi_2}(f_2),$$

using the Riesz decomposition property for F. Therefore, $\rho_{\varphi_1,\varphi_2}$ is an additive, positively homogeneous lower bound for φ_1 and φ_2. By definition, $\rho_{\varphi_1,\varphi_2}$ is in fact the greatest lower bound of this type. If $\varphi \in C$, $\ell \in F'$ and $\ell_\varphi \in L_\varphi$ we thus obtain

$$\ell_\varphi \wedge \ell \leq \rho_{\varphi,\ell} \leq \ell \quad \text{on } F_+.$$

Consequently, $\rho_{\varphi,\ell}$ is continuous at 0, which implies that $\rho_{\varphi,\ell} \in C_o$ and $\rho_{\varphi,\ell} = \varphi \wedge \ell$. For each $\varphi_o \in C$ and $f \in F_+$ we conclude

$$\begin{aligned}(\varphi_o + \varphi \wedge \ell)(f) &= \varphi_o(f) + \inf\{\varphi(g) + \ell(f-g) : g \in F_+,\ g \leq f\} \\ &= \inf\{(\varphi_o + \varphi)(g) + (\varphi_o + \ell)(f-g) : g \in F_+,\ g \leq f\} \\ &= \rho_{\varphi_o+\varphi,\ \varphi_o+\ell}(f).\end{aligned}$$

Since $\rho_{\varphi_o+\varphi,\varphi_o+\ell}$ is the greatest additive, positively homogeneous lower bound for $\varphi_o + \varphi$ and $\varphi_o + \ell$, it follows that

$$(\varphi_o + \varphi) \wedge (\varphi_o + \ell) \leq \varphi_o + \varphi \wedge \ell.$$

The converse inequality being obvious, C is a tight imbedding cone for F'.

The maximal element of C is the functional $\bar{\varphi} : F_+ \to \mathbb{R}_\infty$, given by

$$\bar{\varphi}(0) = 0, \quad \bar{\varphi}(f) = +\infty \quad \text{for all } f \in F_+ \setminus \{0\}.$$

Finally, if $L_1, L_2 \subset F'$ are upward directed such that $\varphi := \sup L_1 = \sup L_2$ in C, then

$$\sup_{\ell_1 \in L_1} (\ell_1 \wedge \ell) = \varphi \wedge \ell = \sup_{\ell_2 \in L_2} (\ell_2 \wedge \ell) \quad \text{for each } \ell \in F'.$$

By Theorem 1.4 , C is therefore isomorphic to the sup-completion of F'. ■

3.14 Lemma: Given a vector space E and a normed vector lattice F let $p: E \times F \to \mathbb{R}_\infty$ be a bisublinear functional, satisfying the following conditions:

(i) $\{f \in F : p(e,f) < \infty\} \subset F_+$ for each $e \in E \setminus \{0\}$,

(ii) $f \to p(e,f)$ is l.s.c. and additive for each $e \in E$.

Then there exists a sublinear mapping Φ from E into the sup-completion F'_s (identified with all l.s.c. additive, positively homogeneous functions $\varphi : F_+ \to \mathbb{R}_\infty$) of F' such that

$$\Phi(e)(f) = p(e,f) \quad \text{for all } f \in F_+.$$

Conversely, if $\Phi : E \to F'_s$ is sublinear, the subbilinear functional $p: E \times F \to \mathbb{R}_\infty$ defined by

$$p(e,f) = \begin{cases} \Phi(e)(f), & \text{for } f \in F_+,\ e \in E, \\ 0, & \text{for } e = 0,\ f \in F \setminus F_+ \\ \infty, & \text{for } e \in E \setminus \{0\},\ f \in F \setminus F_+, \end{cases}$$

satisfies the conditions (i) and (ii).

The proof is clear. ■

Obviously, using the notation of Lemma 3.14, an operator $T: E \to F'$ is p-dominated if it is Φ-dominated.

There are many important extension problems for operators, in particular those concerned with norm-preserving, linear extensions, that can be formalized by subbilinear functionals using the techniques developed in this chapter. The resulting subbilinear functionals, however, are not always additive in one variable. Therefore, it is impossible to treat extension problems of this type successfully using vector-valued sublinear mappings into tight imbedding cones of the target vector lattice.

In order to show that the bisublinear functional $p: E \times F \to \mathbb{R}_\infty$ associated with a sublinear mapping $\Phi : E \to F'_s$ is subbilinear on $E \times F_+$ at least for a Dedekind complete Banach lattice F, we need the following two lemmas.

3.15 Lemma: Let F be a normed vector lattice and $\varphi : F_+ \to \mathbb{R}_\infty$ be a positively homogeneous, additive, l.s.c. functional. If $f \in F_+$ is such that $\varphi(f) < \infty$ and (f_n) is a sequence in F_+ satisfying $\lim\limits_{n\to\infty}(f - f_n)^+ = 0$ and $f_n \leq (1+\frac{1}{n})\cdot f$ for all $n \in \mathbb{N}$, then $\lim\limits_{n\to\infty} \varphi(f_n) = \varphi(f)$.

Proof: Note first that $\lim\limits_{n\to\infty} g_n = 0$ for $g_n := (1+\frac{1}{n})f - f \vee f_n$. Using the equality $\lim\limits_{n\to\infty} f_n = f$ we deduce from the lower semi-continuity of φ

$$\liminf_{n\to\infty} \varphi((f - f_n)^+) \geq 0, \qquad \liminf_{n\to\infty} \varphi(f_n) \geq \varphi(f),$$

$$\liminf_{n\to\infty} \varphi(g_n) \geq 0, \qquad \liminf_{n\to\infty} \varphi(f_n \vee f) \geq \varphi(f).$$

Furthermore, we obtain

$$\lim_{n\to\infty} \varphi((1+\tfrac{1}{n})f) = \varphi(f),$$

since $|\varphi((1+\frac{1}{n})\cdot f) - \varphi(f)| = \frac{1}{n}|\varphi(f)|$ for each $n \in \mathbb{N}$.

Consequently,

$$\begin{aligned}
\varphi(f) + \liminf_{n\to\infty} \varphi(g_n) &\leq \liminf_{n\to\infty} \varphi(f_n \vee f) + \liminf_{n\to\infty} \varphi(g_n) \\
&\leq \limsup_{n\to\infty} \varphi(f_n \vee f) + \liminf_{n\to\infty} \varphi(g_n) \\
&\leq \limsup_{n\to\infty} \varphi(f_n \vee f + g_n) \\
&= \lim_{n\to\infty} \varphi((1+\tfrac{1}{n})f) \\
&= \varphi(f),
\end{aligned}$$

which yields $\liminf\limits_{n\to\infty} \varphi(g_n) = 0$ and $\liminf\limits_{n\to\infty} \varphi(f_n \vee f) = \limsup\limits_{n\to\infty} \varphi(f_n \vee f) = \varphi(f)$. Using the equality

$$\varphi(f_n) + \varphi((f - f_n)^+) = \varphi(f_n + (f - f_n)^+) = \varphi(f \vee f_n)$$

we obtain

$$
\begin{aligned}
\varphi(f) &\leq \liminf_{n\to\infty} \varphi(f_n) + \liminf_{n\to\infty} \varphi(f - f_n)^+) \\
&\leq \limsup_{n\to\infty} \varphi(f_n) + \liminf_{n\to\infty} \varphi((f - f_n)^+) \\
&\leq \lim_{n\to\infty} \varphi(f \vee f_n) \\
&= \varphi(f).
\end{aligned}
$$

Therefore, $\liminf_{n\to\infty} \varphi((f - f_n)^+) = 0$ and $\lim_{n\to\infty} \varphi(f_n) = \varphi(f)$. ■

3.16 Lemma: Given a Dedekind complete vector lattice F, let A be a non-empty, finite subset of F_+ and let $a \in F_+$ be such that A is contained in the vector lattice ideal

$$F_a := \{f \in F : \exists \lambda \geq 0 : |f| \leq \lambda a\}$$

generated by a. Then, for each $n \in \mathbb{N}$, there exists a finite-dimensional vector sublattice F_n of F_a and a positive projection $P_n : F_a \to F_n$ such that $a \in F_n$ and

$$f - \frac{1}{n} \cdot a \leq P_n f \leq (1 + \frac{1}{n}) f.$$

Proof: Since F_a is a Dedekind complete AM-space with order unit a, there exists, by Kakutani's theorem, a compact, extremally disconnected space K and a vector lattice isomorphism $f \to \bar{f}$ from F_a onto $C(K)$, such that $\bar{a} = 1$. If $n \in \mathbb{N}$, choose a closed-open neighborhood U_x for each x in K satisfying

$$\sup \bar{f}(U_x) - \inf \bar{f}(U_x) \leq \frac{1}{n} \quad \text{and}$$

$$\bar{f}(x) \leq (1 + \frac{1}{n}) \cdot \inf \bar{f}(U_x) \quad \text{for all } f \in A.$$

Then there is a finite subset $\{x_1, \ldots, x_m\} \subset K$ such that $\bigcup_{i=1}^{m} U_{x_i} = K$. For each $i \in \{1, \ldots, m\}$ the set

$$V_i := U_{x_i} \setminus \bigcup_{j<i} U_{x_j}$$

is closed-open, and the family $(V_i)_{1\leq i\leq m}$ forms a partition of K. Let G be the set of all functions $g \in C(K)$ constant on every set V_i, $1 \leq i \leq m$. Then G is a finite-dimensional vector sublattice of $C(K)$. It follows that $F_n := \{f \in F_a : \bar{f} \in G\}$ is a finite-dimensional vector sublattice of F_a, hence of F, too.

Consider the projection $P_n : F_a \to F_n$ given by

$$\overline{P_n f} = \sum_{i=1}^{m} \bar{f}(x_i) \cdot 1_{V_i} ,$$

where 1_{V_i} denotes the characteristic function of V_i. For each $f \in A$ and $\xi \in K$ there exists $i \in \{1,\ldots,m\}$ such that $\xi \in V_i \subset U_{x_i}$.

Consequently,

$$\bar{f}(\xi) - \frac{\bar{a}(\xi)}{n} \leq \sup \bar{f}(U_{x_i}) - \frac{1}{n} \leq \inf \bar{f}(U_{x_i}) \leq \bar{f}(x_i)$$

$$= \overline{P_n f}(\xi) \leq (1 + \frac{1}{n}) \cdot \inf \bar{f}(U_{x_i}) \leq (1 + \frac{1}{n}) \bar{f}(\xi),$$

which yields

$$f - \frac{1}{n} a \leq P_n f \leq (1 + \frac{1}{n}) f. \quad \blacksquare$$

We are now prepared to prove

<u>3.17 Theorem</u>: Let E be a vector space, F a Dedekind complete Banach lattice and let $p : E \times F \to \mathbb{R}_\infty$ be a bisublinear functional with the following properties

i) $\{f \in F : p(e,f) < \infty\} \subset F_+$ for each $e \in E \setminus \{0\}$,

ii) $f \to p(e,f)$ is additive and l.s.c. on F_+ for each $e \in E$.

Then p is subbilinear on $E \times F_+$.

<u>Proof</u>: Given $(e,f) \in E \times F_+$ let $(e_i,f_i)_{i\in I}$ be a finite family in $E \times F_+$ such that $\sum_i e_i \otimes f_i = e \otimes f$. Suppose that $\sum_i p(e_i,f_i) < p(e,f)$. Then $p(e_i,f_i) < \infty$ for each $i \in I$. Since $p(0,f') = 0$ for all $f' \in F$, we may assume that $e_i \neq 0$ for each $i \in I$.

Consider the vector lattice ideal F_a generated by $a := f + \sum_i f_i$ in F.

By Lemma 3.16, there exists a sequence of finite-dimensional vector sublattices $(F_n)_{n\in\mathbb{N}}$ of F and a sequence (P_n) of positive projections $P_n : F_a \to F_n$ such that

$$f_i - \frac{1}{n} a \leq P_n f_i \leq (1+\frac{1}{n}) \cdot f_i \qquad (i \in I),$$

$$f - \frac{1}{n} a \leq P_n f \leq (1+\frac{1}{n}) \cdot f$$

for each $n \in \mathbb{N}$. Using Lemma 3.15 we conclude $\lim_{n\to\infty} p(e_i, P_n f_i) = p(e_i, f_i)$ and $\lim_{n\to\infty} p(e, P_n f) = p(e,f)$. Choose $n \in \mathbb{N}$ large enough to satisfy the inequality

$$\sum_i p(e_i, P_n f_i) < p(e, P_n f).$$

Since P_n has a linear extension to F we obtain $\sum_i e_i \otimes P_n f_i = e \otimes P_n f$. Let $(u_k)_{k\in\mathbb{N}}$ be a (necessarily finite) positive orthogonal basis of F_n, i.e. a family of pairwise (order-)disjoint, positive elements in $F_n \setminus \{0\}$ such that each positive element $u \in F_n$ is representable as a sum

$$u = \sum_{k\in K} \lambda_k u_k,$$

where $\lambda_k \geq 0$ for all $k \in K$ (see [66], Ch. II, Cor. 1 of Th. 3.9). If $(\sigma_{ik})_{k\in K} \in \mathbb{R}_+^K$ and $(\sigma_k)_{k\in K} \in \mathbb{R}_+^K$ are the uniquely determined families satisfying

$$\sum_k \sigma_{ik} u_k = P_n f_i, \quad i \in I, \quad \text{and}$$

$$\sum_k \sigma_k u_k = P_n f$$

we obtain

$$\sum_k \sigma_k e \otimes u_k = e \otimes (\sum_k \sigma_k u_k) = e \otimes P_n f = \sum_i e_i \otimes (\sum_k \sigma_{ik} u_k)$$

$$= \sum_k (\sum_i \sigma_{ik} e_i) \otimes u_k.$$

The family $(u_k)_{k\in K}$ being linearly independent, it follows that

$$\sum_i \sigma_{ik} e_i = \sigma_k e \quad \text{for all } k \in K.$$

Finally, using the additivity of p in the second variable on F_+, we conclude

$$\sum_i p(e_i, P_n f_i) = \sum_i p(e_i, \sum_k \sigma_{ik} u_k) = \sum_i \sum_k \sigma_{ik} p(e_i, u_k)$$

$$= \sum_k \sum_i p(\sigma_{ik} e_i, u_k) \geq \sum_k p(\sum_i \sigma_{ik} e_i, u_k) = \sum_k p(\sigma_k e, u_k)$$

$$= \sum_k p(e, \sigma_k u_k) = p(e, \sum_k \sigma_k u_k) = p(e, P_n f),$$

contradicting the inequality $p(e, P_n f) > \sum_i p(e_i, P_n f_i)$. ∎

If E and F are real vector spaces and $p : E \times F \to \mathbb{R}_\infty$ is a subbilinear functional, the equality

$$p(e,f) = \sup\{b(e,f) \ : \ b : E \times F \to \mathbb{R} \text{ bilinear, p-dominated}\}$$

may be violated, although it is generally true for <u>real-valued</u> subbilinear forms by Proposition 3.4. More precisely, we deduce from Theorem 2.12 (see also [2], [32]) that

$$t \to \sup\{b^{\otimes}(t) \ : \ b \text{ p-dominated bilinear form on } E \times F\}$$

is identical with the regularization $\tilde{p}^{\otimes}$ of $p^{\otimes}$ with respect to the finest locally convex topology on $E \otimes F$, i.e.

$$\tilde{p}^{\otimes}(t) = \liminf_{t' \to t} p^{\otimes}(t').$$

In general, $\tilde{p}^{\otimes}$ may be different from $p^{\otimes}$ (see 3.24 for a counterexample). Since the finest locally convex topology on $E \otimes F$ is difficult to handle we shall prefer to use the projective topology on $E \otimes F$ whenever possible. The corresponding regularization of p is the upper envelope of all <u>continuous</u> p-dominated bilinear forms on $E \otimes F$ or, equivalently, of all continuous p-dominated operators from E into F'.

In order to prevent the central ideas from getting hidden behind too much technical framework, we shall restrict our attention to <u>normed</u> spaces E and F, rather than arbitrary locally convex spaces. In this case the bisublinear form $q : E \times F \to \mathbb{R}$ defined by $q(e,f) = \|e\| \cdot \|f\|$ is subbilinear by Example 3.7,i and $q^{\otimes}$ is the projective norm on $E \otimes F$. If, in addition, F is a normed vector lattice, then $q^{\otimes}$ satisfies the conditions (i) and (ii) of the following lemma:

3.18 Lemma: Given a vector space E and a vector lattice F, let $(e_i)_{i \in I}$ and $(g_i)_{i \in I}$, $(f_i)_{i \in I}$ be finite families in E and F_+, respectively, such that $f_i \wedge f_j = 0$ for all $i,j \in I$, $i \neq j$, and $g_i \leq f_i$ for all $i \in I$. Furthermore, suppose that the sublinear form $s : E \otimes F \to \mathbb{R}$ satisfies the following conditions:

i) For every operator $T : E \to F^*$ such that $T^{\otimes} \leq s$ the range $T(E)$ is contained in the subspace $F^b \subset F^*$ of all order bounded linear forms on F, i.e. the order dual.

ii) For every band projection P in F^b and every operator $T : E \to F^b$ such that $T^{\otimes} \leq s$ the inequality $(P \circ T)^{\otimes} \leq s$ holds.

Then $s(\sum_i e_i \otimes g_i) \leq s(\sum_i e_i \otimes f_i)$ and

$$s(\sum_i e_i \otimes f_i) = s(\sum_{i \in I \setminus I'} e_i \otimes f_i)$$

where $I' := \{i \in I : s(e_i \otimes f_i) = 0\}$.

Proof: For each $i \in I$, let

$$C_i := \{f' \in F^b : |f'|(f_i) = 0\} \text{ and}$$

$$C_i^{\perp} := \{g' \in F^b : |g'| \wedge |f'| = 0 \text{ for all } f' \in C_i\}.$$

Then $(C_i^{\perp})_{i \in I}$ is a family of pairwise orthogonal bands in F^b (see e.g. [66], p. 79). Given an operator $T : E \to F^b$ such that $T^{\otimes} \leq s$, consider the band B_i generated by $(Te_i)^+$ in F^b for each $i \in I$. If we denote the respective band projections from F^b onto B_i and $C_i^{\perp}$ by P_i and Q_i, then $R_i := Q_i \circ P_i$ is the band projection from F^b onto the band $B_i \cap C_i^{\perp}$ $(i \in I)$. The bands $B_i \cap C_i^{\perp}$, $i \in I$, being pairwise orthogonal, the sum $R_I := \sum_i R_i$

is the band projection onto the band $\bigoplus_{i\in I}(B_i \cap C_i^{\perp})$ and $(R_I \circ T)^{\otimes} \leq s$ by condition (ii). It follows that

$$s(\sum_i e_i \otimes f_i) \geq \sum_{i\in I}(R_I \circ T)(e_i)(f_i) = \sum_{i\in I}\sum_{j\in I} Q_j(P_j(Te_i))(f_i)$$

$$= \sum_{i\in I} P_i(Te_i)(f_i) = \sum_i (Te_i)^+(f_i) \geq \sum_i (Te_i)^+(g_i)$$

$$\geq \sum_{i\in I} Te_i(g_i)$$

using the equality $Q_j(\ell)(f_i) = 0$ whenever $j \neq i$, $\ell \in F^b$.
Since T was an arbitrary operator satisfying $T^{\otimes} \leq s$ we conclude from the Hahn-Banach theorem that

$$s(\sum_i e_i \otimes f_i) \geq s(\sum_i e_i \otimes g_i).$$

In particular, $s(\sum_{i\in I} e_i \otimes f_i) \geq s(\sum_{i\in I\setminus I'} e_i \otimes f_i)$.

Conversely, if $T: E \to F^b$ is an operator such that $T^{\otimes} \leq s$, then

$$s(\sum_{i\in I\setminus I'} e_i \otimes f_i) \geq \sum_{i\in I\setminus I'} Te_i(f_i) \geq \sum_{i\in I} Te_i(f_i).$$

From this inequality and Hahn-Banach's theorem we obtain

$$s(\sum_{i\in I\setminus I'} e_i \otimes f_i) \geq s(\sum_{i\in I} e_i \otimes f_i). \quad \blacksquare$$

If E and F are normed vector spaces, $q^{\otimes}$ is the projective norm on $E \otimes F$ and $p: E \times F \to \mathbb{R}_{\infty}$ is a bisublinear functional, the equality

$$\tilde{p}^{\otimes}(t) = \sup_{\varepsilon>0} \inf_{q^{\otimes}(t')\leq\varepsilon} p^{\otimes}(t-t')$$

defines a l.s.c. functional $\tilde{p}^{\otimes}: E \otimes F \to \mathbb{R} \cup \{-\infty,\infty\}$.
In fact, $\tilde{p}^{\otimes}$ is the regularization of $p^{\otimes}$ whenever $\tilde{p}^{\otimes}(t) > -\infty$ for all $t \in E \otimes F$. In order to simplify the determination of $\tilde{p}^{\otimes}$ let O_{ε}, $\varepsilon > 0$, denote the set of all finite sequences $(e_i, f_i)_{i\in I}$ in $E \times F_+$ such that $f_i \wedge f_j = 0$ for all $i,j \in I$, $i \neq j$, and $q^{\otimes}(\sum_i e_i \otimes f_i) \leq \varepsilon$. With these notations we have the following

3.19 Theorem: If F is a Dedekind complete vector lattice and the bi-sublinear functional $p : E \times F \to \mathbb{R}_\infty$ satisfies the conditions

$$\{f \in F : p(e,f) < \infty\} \subset F_+ \quad \text{for all } e \in E \setminus \{0\} \text{ and}$$

$$f \to p(e,f) \text{ is additive and l.s.c. on } F_+ \text{ for each } e \in E,$$

then

$$\tilde{p}^{\otimes}(0) = \sup_{\varepsilon>0} \inf_{(e_i,f_i)\in O_\varepsilon} \sum_i p(-e_i,f_i).$$

In order that there exists a continuous p-dominated operator $T : E \to F'$, the condition $\tilde{p}^{\otimes}(0) > -\infty$ is necessary and sufficient. In this case, the equality $\tilde{p}^{\otimes}(0) = 0$ must hold and

$$\tilde{p}^{\otimes}(e \otimes f) = \sup_{\varepsilon>0} \inf_{(e_i,f_i)\in O_{\varepsilon,f}} \sum_i p(e - e_i,f_i)$$

for each $(e,f) \in E \times F_+$, where $O_{\varepsilon,f} := \{(e_i,f_i) \in O_\varepsilon : \sum_i f_i = f\}$.

Proof: Given $\varepsilon > 0$ we obtain the following equalities for each $e \in E$, $f \in F_+$:

$$\inf_{q^{\otimes}(t)\leq\varepsilon} p^{\otimes}(e \otimes f - t) = \inf_{q^{\otimes}(e\otimes f-t')\leq\varepsilon} p^{\otimes}(t') =$$

$$= \inf\{\sum_i p(e_i,f_i) : (e_i,f_i) \text{ finite sequence in } E \times F, q^{\otimes}(e \otimes f - \sum_i e_i \otimes f_i) \leq \varepsilon\}$$

$$= \inf\{\sum_i p(e_i,f_i) : (e_i,f_i) \text{ finite sequence in } E\times F_+, q^{\otimes}(e \otimes f - \sum_i e_i \otimes f_i) \leq \varepsilon\},$$

since $p(e_i,f_i) = +\infty$, whenever $e_i \neq 0$, $f_i \in F \setminus F_+$.
In particular,

$$\inf_{q^{\otimes}(t)\leq\varepsilon} p^{\otimes}(-t) = \inf\{\sum_i p(-e_i,f_i) : (e_i,f_i) \text{ fin. sequ. in } E \times F_+ \text{ such that } q^{\otimes}(\sum_i e_i \otimes f_i) \leq \varepsilon\}.$$

Consider a finite sequence (e_i,f_i) in $E \times F_+$ satisfying $q^{\otimes}(\sum_i e_i \otimes f_i) \leq \varepsilon$. By Lemma 3.15 and 3.16 there is a finite-dimensional vector sublattice U of F_a, $a := \sum_i f_i$, and elements $g_i := Pf_i \in F_a$ $(i \in I)$, P being a suitable projection from F_a onto U such that

$$\sum_i p(-e_i,g_i) \leq \sum_i p(-e_i,f_i) + \varepsilon \quad \text{and} \quad \sum_i \|e_i\| \cdot \|f_i - g_i\| \leq \varepsilon.$$

If we select a positive orthogonal basis $(u_k)_{k\in K}$ of U, then, for each $i \in I$, there exists a uniquely determined family $(\lambda_{ik})_{k\in K}$ of non-negative real numbers such that

$$g_i = \sum_k \lambda_{ik} u_k.$$

The function $f \to p(e,f)$ being additive on F_+ for each $e \in E$ it follows that

$$\varepsilon + \sum_i p(-e_i,f_i) \geq \sum_i p(-e_i,g_i) = \sum_i p(-e_i, \sum_k \lambda_{ik} u_k) = \sum_i \sum_k \lambda_{ik} p(-e_i,u_k)$$

$$\geq \sum_k p(-\sum_i \lambda_{ik} e_i, u_k) = \sum_k p(-e'_k,u_k),$$

where $e'_k := \sum_i \lambda_{ik} e_i$ for each $k \in K$.

From the equality $\sum_k e'_k \otimes u_k = \sum_i e_i \otimes g_i$ we deduce

$$q^{\otimes}(\sum_k e'_k \otimes u_k) = q^{\otimes}(\sum_i e_i \otimes g_i) \leq q^{\otimes}(\sum_i e_i \otimes f_i) + q^{\otimes}(\sum_i e_i \otimes (g_i - f_i))$$

$$\leq \varepsilon + \sum_i \|e_i\| \cdot \|g_i - f_i\| \leq 2\varepsilon,$$

hence $(e'_k,u_k) \in O_{2\varepsilon}$, which yields

$$\sum_i p(-e_i,f_i) \geq \inf_{(e''_\ell,f''_\ell)\in O_{2\varepsilon}} \sum_\ell p(-e''_\ell,f''_\ell) - \varepsilon.$$

Since $\sum_i e_i \otimes f_i$ was an arbitrary tensor satisfying $q^{\otimes}(\sum_i e_i \otimes f_i) \leq \varepsilon$, we obtain

$$\tilde{p}^{\otimes}(0) \geq \inf_{q^{\otimes}(t)\leq\varepsilon} p^{\otimes}(-t) \geq \inf_{(e''_\ell,f''_\ell)\in O_{2\varepsilon}} \sum_\ell p(-e''_\ell,f''_\ell) - \varepsilon.$$

Letting ε tend to 0, it follows that

$$\tilde{p}^{\otimes}(0) \geq \sup_{\varepsilon>0} \inf_{(e''_\ell,f''_\ell)\in O_\varepsilon} \sum_\ell p(-e''_\ell,f''_\ell).$$

The converse inequality being evident, we conclude

$$\tilde{p}^{\otimes}(0) = \sup_{\varepsilon>0} \inf_{(e''_\ell,f''_\ell)\in O_\varepsilon} \sum_\ell p(-e''_\ell,f''_\ell).$$

Since, clearly, $\alpha\tilde{p}^{\otimes}(0) = \tilde{p}^{\otimes}(0)$ for all $\alpha > 0$ we must have $\tilde{p}^{\otimes}(0) = 0$ provided that $\tilde{p}^{\otimes}(0) > -\infty$. Under this condition $p^{\otimes}$ is therefore l.s.c. at 0. By Proposition 2.3 there exists a $p^{\otimes}$-dominated, continuous linear form ℓ on $E \otimes F$. The associated operator $T : E \to F'$, given by

$$Te(f) := \ell(e \otimes f)$$

is continuous and p-dominated.

In order to complete the proof assume that $p^{\otimes}$ is l.s.c. at 0 and that $T_o : E \to F'$ is a continuous, p-dominated operator. Given $e \in E$, $f \in F_+$, we claim that

$$\inf\{\sum_i p(e_i,f_i) : (e_i,f_i) \text{ finite sequence in } E \times F_+,\ q^{\otimes}(e \otimes f - \sum_i e_i \otimes f_i) \leq \varepsilon\}$$

$$\geq \inf_{(e_i,f_i) \in O_{3\varepsilon,f}} \sum_i p(e - e_i,f_i) - (2\|T_o\| + 1) \cdot \varepsilon$$

for all $\varepsilon > 0$. From this inequality the relation

$$\tilde{p}^{\otimes}(e \otimes f) = \sup_{\varepsilon > 0} \inf_{(e_i,f_i) \in O_{\varepsilon,f}} \sum_i p(e - e_i,f_i)$$

will immediately follow. To prove the inequality, let $\varepsilon > 0$ and a finite sequence (e_i,f_i) in $E \times F_+$ be given such that $q^{\otimes}(e \otimes f - \sum_i e_i \otimes f_i) \leq \varepsilon$. For each $n \in \mathbb{N}$, consider the band projection P_n from F onto the band orthogonal to $(nf - b)^-$, where $b := f + \sum_i f_i$. From the inequality $P_n(nf-b) = (nf - b)^+ \geq 0$ we deduce that

$$P_n f_i \leq P_n b \leq n \cdot P_n f \quad \text{for all } i \in I \quad \text{and}$$

$$P_n f - f = (f - \frac{1}{n} b)^+ + \frac{1}{n} P_n b - f$$

hence $\lim_{n\to\infty} P_n f = f$. Choose n large enough to satisfy $\|e\| \cdot \|P_n f - f\| \leq \varepsilon$. If Id denotes the identity mapping on F, then $Id - P_n$ is the band projection orthogonal to P_n. In particular, $Id - P_n$ and its adjoint $(Id - P_n)'$ are contractive operators. Therefore

$$\sum_i ((Id - P_n)' \circ T_o)(-e_i)(f_i) + ((Id - P_n)' \circ T_o)(e)(f) \leq$$

$$\leq \|(Id - P_n)'\circ T_o\| q^{\otimes}(-\sum_i e_i \otimes f_i + e \otimes f) \leq \|T_o\| \varepsilon,$$

which yields

$$\begin{aligned}\sum_i p(e_i, f_i - P_n f_i) &\geq \sum_i T_o e_i (f_i - P_n f_i) = \sum_i ((Id - P_n)'\circ T_o)(e_i)(f_i) \\ &\geq -\varepsilon\|T_o\| - ((Id - P_n)'\circ T_o)(e)(f) \\ &= -\varepsilon\|T_o\| - T_o e(f - P_n f) \\ &\geq -\varepsilon\|T_o\| - \|T_o\| \cdot \|e\| \cdot \|f - P_n f\| \geq -2\varepsilon\|T_o\|.\end{aligned}$$

Since, for each $i \in I$, $P_n f_i$ is a member of the vector lattice ideal V generated by $P_n f$ in F, an application of Lemma 3.15 and 3.16 ensures the existence of a finite dimensional vector sublattice U of V and an associate positive projection $Q : V \to U$ such that

$$P_n f \in U, \quad \sum_i \|e_i\| \cdot \|Q(P_n f_i) - P_n f_i\| \leq \varepsilon \quad \text{and}$$

$$\sum_i p(e_i, Q(P_n f_i)) \leq \sum_i p(e_i, P_n f_i) + \varepsilon$$

$P_n f$ being an order unit in U, there exists a positive orthogonal basis $(u_k)_{1\leq k\leq m}$ of U such that $\sum_{k=1}^{m} u_k = P_n f$ (see [66], Ch. II, Cor. 1 of Th. 3.9). Let $(\lambda_{ik})_{1\leq k\leq m}$ denote the uniquely determined family of non-negative real numbers satisfying $Q(P_n f_i) = \sum_{k=1}^{m} \lambda_{ik} u_k$, $i \in I$. If we set $e'_o := e$, $u_o := f - P_n f$ and $e'_k := e - \sum_i \lambda_{ik} e_i$ for each $k \in \{1,\dots,m\}$, then we deduce from the additivity of p with respect to the second variable on F_+:

$$2\varepsilon\|T_o\| + \varepsilon + \sum_i p(e_i, f_i) = 2\varepsilon\|T_o\| + \varepsilon + \sum_i p(e_i, f_i - P_n f_i) + \sum_i p(e_i, P_n f_i) \geq$$

$$\geq \sum_i p(e_i, Q(P_n f_i)) = \sum_i p(e_i, \sum_{k=1}^{m} \lambda_{ik} u_k) = \sum_i \sum_k \lambda_{ik} p(e_i, u_k) \geq$$

$$\geq \sum_k p(\sum_i \lambda_{ik} e_i, u_k) = \sum_k p(e - e'_k, u_k) + p(e - e'_o, u_o).$$

In order to show that $(e'_k, u_k)_{0\leq k\leq m} \in O_{3\varepsilon,f}$ note that

$$\sum_{k=o}^{m} u_k = f - P_n f + \sum_{k=1}^{m} u_k = f,$$

$u_k \wedge u_{k'} = 0$ for $0 \leq k < k' \leq m$

and

$$q^{\otimes}(\sum_{k=o}^{m} e'_k \otimes u_k) \leq q^{\otimes}(\sum_{k=1}^{m} (e - \sum_i \lambda_{ik} e_i) \otimes u_k) + q^{\otimes}(e \otimes (f - P_n f)) \leq$$

$$\leq q^{\otimes}(e \otimes P_n f - \sum_i e_i \otimes Q(P_n f_i)) + \|e\| \cdot \|f - P_n f\| \leq$$

$$\leq q^{\otimes}(e \otimes P_n f - \sum_i e_i \otimes P_n f_i) + q^{\otimes}(\sum_i e_i \otimes (P_n f_i - Q(P_n f_i))) + \varepsilon \leq$$

$$\leq \sup\{Te(P_n f) - \sum_i Te_i(P_n f_i) : T : E \to F' \text{ contractive}\} + \sum_i \|e_i\| \, \|Q(P_n f_i) - P_n f_i\| + \varepsilon$$

$$\leq \sup\{(P'_n \circ T)e(f) - \sum_i (P'_n \circ T)e_i(f_i) : T : E \to F' \text{ contractive}\} + 2\varepsilon$$

$$\leq q^{\otimes}(e \otimes f - \sum_i e_i \otimes f_i) + 2\varepsilon \leq 3\varepsilon.$$

We thus conclude

$$\sum_i p(e_i, f_i) \geq \inf_{(e''_j, f''_j) \in O_{3\varepsilon, f}} \sum_j p(e''_j, f''_j) - (2\|T_o\| + 1) \cdot \varepsilon,$$

which completes the proof. ■

The simpler description of the l.s.c. functional $\tilde{p}^{\otimes}$ obtained in Theorem 3.19 will soon prove rather useful. In order to modify the result such that extensions of <u>positive</u> operators are included we need the following bisublinear functional $r : E \times F \to \mathbb{R}_{\infty}$ for normed vector lattices E and F:

$$r(e,f) = \begin{cases} \|e^+\| \cdot \|f\|, & \text{for } e \in E,\ f \in F_+ \\ 0 & \text{if } e = 0,\ f \in F \setminus F_+ \\ \infty & \text{else.} \end{cases}$$

By Proposition 3.4, $r^{\otimes}$ is a real-valued sublinear form on $E \otimes F$. It is easy to see that an operator $T : E \to F^*$ is r-dominated iff it is positive (which implies, in particular, $T(E) \subset F^b = F'$) and contractive as an operator from E into the topological dual F' of F.

Indeed, if $T : E \to F'$ is a positive contraction, the inequality

$$Te(f) \leq Te^+(f) \leq \|e^+\| \cdot \|f\| = r(e,f)$$

holds for each pair $(e,f) \in E \times F_+$.

Conversely, suppose that a given operator $T : E \to F^*$ satisfies this inequality. Then

$$Te(f) \leq \|e\| \cdot \|f\| \text{ for each } (e,f) \in E_+ \times F_+.$$

Since $F = F_+ - F_+$ we deduce that $T(E_+) \subset F'$ and hence

$$T(E) = T(E_+) - T(E_+) \subset F'.$$

Moreover, from the inequality

$$T(-e)(f) \leq \|(-e^+)\| \cdot \|f\| = 0 \quad (f \in F_+)$$

we conclude $T(-e) \leq 0$ for all $e \in E_+$ which implies that T is positive.

Finally, T is also contractive, since

$$\|Te\| \leq \|T(|e|)\| = \sup_{\substack{f \in F_+ \\ \|f\| \leq 1}} T(|e|)(f) \leq \sup_{\substack{f \in F_+ \\ \|f\| \leq 1}} \|e\| \cdot \|f\| = \|e\| \quad (e \in E).$$

Using Proposition 3.4 we obtain

$r^{\otimes}(t) = \sup\{T^{\otimes}(t) : T \text{ positive contraction from } E \text{ into } F'\}$.

Furthermore, $r^{\otimes}$ obviously satisfies the conditions (i) and (ii) of Lemma 3.12, and, for every positive continuous operator $T : E \to F'$ we have

$$T^{\otimes}(t) \leq \|T\| \cdot r^{\otimes}(t) \quad (t \in E \otimes F).$$

<u>3.20 Corollary of Theorem 3.19</u>: Given a normed vector lattice E and a Dedekind complete Banach lattice F, let $p : E \times F \to \mathbb{R}_\infty$ be a bisublinear functional satisfying the conditions (i), (ii) of 3.17. Assume that the functional $e \to p(e,f)$ is increasing for each $f \in F_+$. Let, for $\varepsilon > 0$, O_ε^+ denote the set of all finite sequences $(e_i, f_i)_{i \in I}$ in $E \times F_+$ such

that $r^{\otimes}(\sum_i e_i \otimes f_i) \leq \varepsilon$ and $f_i \wedge f_j = 0$ whenever $i,j \in I$, $i \neq j$. Then

$$\widetilde{p}^{\otimes}(0) = \sup_{\varepsilon>0} \inf_{(e_i,f_i)\in O_\varepsilon^+} \sum_i p(-e_i,f_i)$$

and, if $\widetilde{p}^{\otimes}(0) > -\infty$,

$$\widetilde{p}^{\otimes}(e \otimes f) = \sup_{\varepsilon>0} \inf_{(e_i,f_i)\in O_{\varepsilon,f}^+} \sum_i p(e - e_i,f_i) \quad \text{for each pair } (e,f) \in E \times F_+,$$

where $O_{\varepsilon,f}^+ = \{(e_i,f_i)_{i\in I} \in O_\varepsilon^+ : \sum_i f_i = f\}$.

Proof: By Theorem 3.19 it suffices to show that

$$\sup_{\varepsilon>0} \inf_{(e_i,f_i)\in O_\varepsilon^+} \sum_i p(-e_i,f_i) = \sup_{\varepsilon>0} \inf_{(e_i,f_i)\in O_\varepsilon} \sum_i p(-e_i,f_i) \quad \text{and}$$

$$\sup_{\varepsilon>0} \inf_{(e_i,f_i)\in O_{\varepsilon,f}^+} \sum_i p(e - e_i,f_i) = \sup_{\varepsilon>0} \inf_{(e_i,f_i)\in O_{\varepsilon,f}} \sum_i p(e - e_i,f_i)$$

for all $(e,f) \in E \times F_+$.

Using the obvious inequality $q^{\otimes} \geq r^{\otimes}$ we obtain the inclusions $O_\varepsilon \subset O_\varepsilon^+$ and $O_{\varepsilon,f} \subset O_{\varepsilon,f}^+$. Therefore it remains to check the inequalities

$$\sup_{\varepsilon>0} \inf_{(e_i,f_i)\in O_\varepsilon^+} \sum_i p(-e_i,f_i) \geq \widetilde{p}^{\otimes}(0) \quad \text{and}$$

$$\sup_{\varepsilon>0} \inf_{(e_i,f_i)\in O_{\varepsilon,f}^+} \sum_i p(e - e_i,f_i) \geq \widetilde{p}^{\otimes}(e \otimes f) \quad \text{for } (e,f) \in E \times F_+.$$

If $\widetilde{p}^{\otimes}(0) = -\infty$, the first inequality obviously holds. Hence we may assume $\widetilde{p}^{\otimes}(0) > -\infty$, which is equivalent to $\widetilde{p}^{\otimes}(0) = 0$ by Theorem 3.19. Consider a continuous p-dominated operator $T : E \to F'$. The function $e' \to p(e',f')$ being isotone, provided that $f' \in F_+$, we have

$$T(-e')(f') \leq p(-e',f') \leq p(0,f') = 0 \quad \text{for each pair } (e',f') \in E_+ \times F_+$$

which implies $T(-e') \leq 0$. Therefore, T is positive.

Consequently, if $\varepsilon > 0$ and $(e_i,f_i) \in O_\varepsilon^+$ are given, we obtain

$$\sum_i Te_i(f_i) \leq \|T\| r^{\otimes}(\sum_i e_i \otimes f_i) \leq \|T\| \cdot \varepsilon \quad \text{and}$$

$$-\varepsilon\|T\| \leq \sum_i T(-e_i)(f_i) \leq \sum_i p(-e_i,f_i).$$

It follows that

$$-\varepsilon\|T\| \leq \sup_{\varepsilon'>0} \inf_{(e_i,f_i)\in O^+_{\varepsilon'}} \sum_i p(-e_i,f_i),$$

and, $\varepsilon > 0$ being arbitrary,

$$\tilde{p}^{\otimes}(0) = 0 \leq \sup_{\varepsilon>0} \inf_{(e_i,f_i)\in O^+_{\varepsilon}} \sum_i p(-e_i,f_i).$$

The inequality $\tilde{p}^{\otimes}(e\otimes f) \leq \sup_{\varepsilon>0} \inf_{(e_i,f_i)\in O^+_{\varepsilon,f}} \sum_i p(e-e_i,f_i)$,

$(e,f) \in E\times F_+$, is proved similarly. ∎

3.21 Example: Let E,F,G be normed vector lattices such that F' is a normed vector sublattice of G. Consider a positive, continuous operator T from a linear subspace H of E into F' possessing a positive continuous linear extension $T_o : E \to G$. For each $\varepsilon > 0$ and $e \in E$ let $H_{e,\varepsilon}$ denote the set of all finite sequences (h_i) in H such that

$$\|(e - \bigwedge_{i\in I} h_i)^+\| \leq \varepsilon. \quad ^{+)}$$

If $e \in E$, $f \in F_+$, $\varepsilon > 0$, and $\bar{f}$ is a positive, norm-preserving, linear extension of $\ell \to \ell(f)$ $(\ell \in F')$ to G, then, for each $(h_i) \in H_{e,\varepsilon}$

$$(\bigwedge_i Th_i)(f) = \bar{f}(\bigwedge_i T_oh_i) \geq \bar{f}(T_o(\bigwedge_i h_i)) = \bar{f}(T_oe) - \bar{f}(T_o(e - \bigwedge_i h_i))$$

$$\geq \bar{f}(T_oe) - \|T_o\|\cdot\|(e - \bigwedge_i h_i)^+\|\cdot\|f\| \geq \bar{f}(T_oe) - \varepsilon\|T_o\|\cdot\|f\|.$$

Consequently,

$$\sup_{\varepsilon>0} \inf_{(h_i)\in H_{e,\varepsilon}} (\bigwedge_i Th_i)(f) > -\infty \quad \text{for each } e \in E,\ f \in F_+,$$

where we use the convention $\inf \emptyset := +\infty$. The function $\hat{T} : E \times F \to \mathbb{R}_\infty$

+) $\bigwedge_{i\in I} h_i := \inf\{h_i : i\in I\}$, $\bigvee_{i\in I} h_i := \sup\{h_i : i\in I\}$, respectively.

defined by

$$\hat{T}(e,f) = \begin{cases} \sup\limits_{\varepsilon>0} \inf\limits_{(h_i)\in H_{e,\varepsilon}} (\bigwedge\limits_i Th_i)(f) & \text{for } e\in E,\ f\in F_+, \\ 0, & \text{whenever } e = 0,\ f\in F\setminus F_+, \\ \infty, & \text{if } e\in E\setminus\{0\},\ f\in F\setminus F_+, \end{cases}$$

is in fact a bisublinear functional satisfying all assumptions of Corollary 3.20 as we shall show in the following lemma. Moreover, in 3.23 it is proved that an operator $T_1 : E \to G$ is a continuous, positive, linear extension of $T : H \to F'$ iff T_1 is $\hat{T}$-dominated.

<u>3.22 Lemma</u>: The functional $\hat{T}$ is bisublinear and has the following properties:

i) $\{f\in F : \hat{T}(e,f) < \infty\} \subset F_+$ for all $e\in E\setminus\{0\}$,

ii) $f \to \hat{T}(e,f)$ is additive on F_+ for each $e\in E$,

iii) $e \to \hat{T}(e,f)$ is isotone for each $f\in F_+$.

In particular, $\hat{T}$ is subbilinear on $E\times F_+$, provided that F is Dedekind complete. Furthermore, $\hat{T}$ is l.s.c. with respect to the product topology on $E\times F$ and $\hat{T}(h,f) = Th(f)$ for each $h\in H$, $f\in F_+$.

<u>Proof</u>: Given $e\in E$, let us first show that $f\to\hat{T}(e,f)$ is sublinear. Since the positive homogenity is obvious and since $\hat{T}(e,f_1+f_2) \leq \hat{T}(e,f_1)+\hat{T}(e,f_2)$ whenever $f_1\notin F_+$ or $f_2\notin F_+$, it suffices to prove the subadditivity on F_+. Let $f_1,f_2\in F_+$, $\varepsilon>0$ and $(h_i)_{1\leq i\leq m}$, $(k_i)_{1\leq i\leq n} \in H_{e,\varepsilon/2}$; then $(h_i')_{1\leq i\leq m+n}\in H_{e,\varepsilon}$, where

$$h_i' := \begin{cases} h_i, & \text{if } i\in\{1,\dots,m\} \\ k_{i-m}, & \text{if } i\in\{m+1,\dots,m+n\}. \end{cases}$$

Consequently,

$$\inf_{(h_i')\in H_{e,\varepsilon}} (\bigwedge_i Th_i')(f_1+f_2) \leq \inf_{(h_i)\in H_{e,\varepsilon/2}} (\bigwedge_i Th_i)(f_1) + \inf_{(k_i)\in H_{e,\varepsilon/2}} (\bigwedge_i Tk_i)(f_2)$$

which implies

$$\inf_{(h_i')\in H_{e,\varepsilon}} (\bigwedge_i Th_i')(f_1+f_2) \le \hat{T}(e,f_1)+\hat{T}(e,f_2).$$

Since $\varepsilon > 0$ was arbitrary, we conclude

$$\hat{T}(e,f_1+f_2) \le \hat{T}(e,f_1)+\hat{T}(e,f_2).$$

To prove that $e \to \hat{T}(e,f)$ is sublinear for $f \in F$ it suffices to check the inequality $\hat{T}(e_1+e_2,f) \le \hat{T}(e_1,f)+\hat{T}(e_2,f)$ for $e_1,e_2 \in E$ and $f \in F_+$, the other cases and the positive homogenity being obvious.
Given $\varepsilon > 0$ and $(h_i)_{i\in I} \in H_{e_1,\varepsilon/2}$, $(k_j)_{j\in J} \in H_{e_2,\varepsilon/2}$ set $h_{ij} := h_i + k_j$ for each pair $(i,j) \in I \times J$. Then the relation

$$\|(e_1+e_2-\bigwedge_{(i,j)\in I\times J} h_{ij})^+\| = \|(e_1+e_2-(\bigwedge_i h_i + \bigwedge_j k_j))^+\| \le$$

$$\le \|(e_1-\bigwedge_i h_i)^+\| + \|(e_2-\bigwedge_j k_j)^+\| \le \varepsilon$$

implies that (h_{ij}), rewritten as a sequence, lies in $H_{e_1+e_2,\varepsilon}$.
Therefore

$$\inf_{(h_\ell')\in H_{e_1+e_2,\varepsilon}} (\bigwedge_\ell Th_\ell')(f) \le \inf_{(h_i)\in H_{e_1,\varepsilon/2}} (\bigwedge_i Th_i)(f) + \inf_{(k_j)\in H_{e_2,\varepsilon/2}} (\bigwedge_j Tk_j)(f)$$

$$\le \hat{T}(e_1,f)+\hat{T}(e_2,f),$$

which yields $\hat{T}(e_1+e_2,f) \le \hat{T}(e_1,f)+\hat{T}(e_2,f)$.
Since property (i) is evident, let $e \in E$ be given in order to show that $f \to \hat{T}(e,f)$ is additive on F_+. If $f_1,f_2 \in F_+$ and $H_{e,\varepsilon} = \emptyset$ for some $\varepsilon > 0$ then $\hat{T}(e,f_1+f_2) = \infty = \hat{T}(e,f_1)+\hat{T}(e,f_2)$. Hence we may assume that $H_{e,\varepsilon} \neq \emptyset$ for every $\varepsilon > 0$. Selecting $(k_j) \in H_{e,\varepsilon}$ such that

$$(\bigwedge_j Tk_j)(f_1+f_2) \le \inf_{(h_i)\in H_{e,\varepsilon}} (\bigwedge_i Th_i)(f_1+f_2)+\varepsilon \le \hat{T}(e,f_1+f_2)+\varepsilon$$

we obtain

$$\inf_{(h_i)\in H_{e,\varepsilon}} (\bigwedge_i Th_i)(f_1) + \inf_{(h_i)\in H_{e,\varepsilon}} (\bigwedge_i Th_i)(f_2) \leq (\bigwedge_j Tk_j)(f_1) + (\bigwedge_j Tk_j)(f_2)$$
$$\leq \hat{T}(e,f_1+f_2)+\varepsilon.$$

Since $\varepsilon > 0$ was arbitrary, it follows from the subadditivity of $f \to \hat{T}(e,f)$ that

$$\hat{T}(e,f_1)+\hat{T}(e,f_2) = \hat{T}(e,f_1+f_2).$$

Clearly, $e \to \hat{T}(e,f)$ is increasing for $f \in F_+$. Furthermore, given $h \in H$, $f \in E_+$ and $\varepsilon > 0$ we obtain the inequality

$$Th(f) - \varepsilon\|T_o\|\cdot\|f\| \leq Th(f) - \bar{f}(T_o((h - \bigwedge_i h_i)^+)) \leq (\bigwedge_i Th_i)(f)$$

for each family $(h_i) \in H_{h,\varepsilon}$, $\bar{f}$ denoting a positive, norm-preserving linear extension of $\ell \to \ell(f)$, $\ell \in F'$, to G. Since h is a member of $\{h' \in H : \|(h-h')^+\| \leq \varepsilon\}$ we deduce that

$$Th(f) - \varepsilon\|T_o\|\cdot\|f\| \leq \inf_{(h_i)\in H_{h,\varepsilon}} (\bigwedge_i Th_i)(f) \leq Th(f).$$

Letting ε tend to 0 we finally get $Th(f) = \hat{T}(h,f)$.

To complete the proof it remains to show that $\hat{T}$ is l.s.c. at each point $(e,f) \in E \times F$. If $f \in F \setminus F_+$, then $\hat{T}$ is clearly l.s.c. at (e,f) since F_+ is closed in F. Hence we may assume that $f \in F_+$. Given a real number $\alpha < \hat{T}(e,f)$ and $\varepsilon > 0$ choose $\varepsilon_1 > 0$ such that

$\inf_{(h_i)\in H_{e,3\varepsilon_1}} (\bigwedge_i Th_i)(f) > \alpha$. If $(k_i)_{1\leq i\leq m} \in H_{e,\varepsilon_1}$, there exists a positive real number ε_2 satisfying the following conditions:

$3\|T_o\|\varepsilon_1\varepsilon_2 < \varepsilon$,

$\|(\bigwedge_i T_o(k_i - e))^+\|\cdot\varepsilon_2 < \varepsilon$,

$\|T_o e\|\cdot\varepsilon_2 \leq \varepsilon$.

We claim that

(3.22.1) $\hat{T}(e',f') \geq \alpha - 3\varepsilon$

for all $(e',f') \in E \times F_+$ such that $\|e-e'\| \leq \varepsilon_1$, $\|f-f'\| \leq \varepsilon_2$.

From this inequality the lower semicontinuity at (e,f) follows.

Indeed, if $e \neq 0$, we deduce from (3.22.1):

$\hat{T}(e',f') \geq \alpha - 3\varepsilon$

for all $(e',f') \in E \times F$ satisfying $\|e - e'\| < \min(\varepsilon_1, \|e\|), \|f - f'\| \leq \varepsilon_2$, since $\hat{T}(e',f') = \infty > \alpha - 3\varepsilon$ whenever $f' \in F \setminus F_+$, $e' \neq 0$.

If $e = 0$, (3.22.1) yields

$\hat{T}(e',f') \geq \alpha - 3\varepsilon$

for all $(e',f') \in E \times F$ satisfying $\|e - e'\| \leq \varepsilon_1, \|f - f'\| \leq \varepsilon_2$,

using the inequalities

$$\alpha - 3\varepsilon < \infty = \hat{T}(e',f') \text{ valid for } e' \neq 0,\ f' \in F \setminus F_+ \text{ and}$$

$$\alpha - 3\varepsilon < \hat{T}(e,f) = 0 = \hat{T}(e',f') \text{ whenever } e' = 0,\ f' \in F \setminus F_+.$$

In order to prove (3.22.1), let $e' \in E$, $\|e' - e\| \leq \varepsilon_1$, and $(h_i)_{1 \leq i \leq n} \in H_{e',\varepsilon_1}$ be given. From the relation

$$0 \leq (e - \bigwedge_i h_i)^+ = (e - e' + e' - \bigwedge_i h_i)^+ \leq (e - e')^+ + (e' - \bigwedge_i h_i)^+$$

we deduce $\|(e - \bigwedge_i h_i)^+\| \leq 2\varepsilon_1$, i.e. $(h_i) \in H_{e,2\varepsilon_1}$.

If we set $h_i' := h_i$ for $1 \leq i \leq n$ and $h_i' := k_{i-n}$ for $n < i \leq n+m$, then

$$\|(e - \bigwedge_{i=1}^{m+m} h_i')^+\| = \|e - \bigwedge_{i=1}^{n} h_i)^+ \vee (e - \bigwedge_{i=1}^{m} k_i)^+\| \leq 3\varepsilon_1,$$

i.e. $(h_i')_{1 \leq i \leq n+m} \in H_{e,3\varepsilon_1}$.

Hence, if $f' \in F_+$ is such that $\|f' - f\| \leq \varepsilon_2$ and if $\varphi \in G'$ is a norm-preserving positive extension of $\ell \to \ell(|f - f'|)$, $\ell \in F'$, to G, then the following inequalities hold:

$$|(\bigwedge_{i=1}^{m+n} Th_i')(f'-f)| \leq |\bigwedge_{i=1}^{m+n} Th_i'|(|f'-f|) \leq \varphi(|\bigwedge_{i=1}^{m+n} T_o(h_i'-e)|) + \varphi(|T_o e|)$$

$$\leq \varphi((\bigwedge_{i=1}^{m+n} T_o(h_i'-e))^+) + \varphi((\bigvee_{i=1}^{m+n} T_o(e-h_i'))^+) + \varepsilon_2 \cdot \|T_o e\|$$

$$\leq \varphi((\bigwedge_{i=1}^{m} T_o(k_i-e))^+) + \varphi(T_o(\bigvee_{i=1}^{m+n} (e-h_i')^+)) + \varepsilon$$

$$\leq \| \bigwedge_{i=1}^{m} T_o(k_i - e))^+ \| \cdot \varepsilon_2 + \varphi(T_o((e - \bigwedge_{i=1}^{m+n} h_i')^+)) + \varepsilon$$

$$\leq 2\varepsilon + \|T_o\| \cdot 3\varepsilon_1 \cdot \varepsilon_2 \leq 3\varepsilon.$$

Consequently,

$$(\bigwedge_{i=1}^{n} Th_i)(f') \geq (\bigwedge_{i=1}^{m+n} Th_i')(f') \geq (\bigwedge_{i=1}^{m+n} Th_i')(f) - 3\varepsilon$$

$$\geq \inf_{(h_j'') \in H_{e,3\varepsilon_1}} (\bigwedge_j Th_j'')(f) - 3\varepsilon > \alpha - 3\varepsilon.$$

Since $(h_i) \in H_{e';\varepsilon_1}$ was arbitrary, this yields

$$\hat{T}(e',f') \geq \inf_{(h_i) \in H_{e',\varepsilon_1}} (\bigwedge_i Th_i)(f') > \alpha - 3\varepsilon$$

completing the proof. ■

Although the functional $\hat{T}$ seems to be rather complicated at first sight it is of central importance for the subsequent sections. This is pointed out by the following

3.23 Lemma: Let E,F,G be vector lattices such that the topological dual F' of F is a normed vector sublattice of G. Furthermore, let T be a positive continuous operator defined on some linear subspace H of E into F' such that there exists a positive, continuous, linear extension $T_o : E \to G$. Then every positive, continuous, linear extension $T_1 : E \to G$ of T is $\hat{T}$-dominated. Conversely, every $\hat{T}$-dominated operator $T_1 : E \to G$ is a positive extension of T.

Proof: Given a positive, continuous, linear extension $T_1 : E \to F'$ of T, $(e,f) \in E \times F_+$ and $\varepsilon > 0$, choose $\varepsilon' > 0$ such that $\varepsilon' \cdot \|T_1\| \cdot \|f\| \leq \varepsilon$. For every finite sequence $(h_i) \in H_{e,\varepsilon'}$, we obtain:

$$T_1e(f) - \varepsilon \leq (\bigwedge_i T_1h_i)(f) + (\bigvee_i T_1(e-h_i))^+(f) - \varepsilon \leq$$

$$\leq (\bigwedge_i Th_i)(f) + T_1(\bigvee_i (e-h_i)^+(f) - \varepsilon \leq (\bigwedge_i Th_i)(f),$$

since $\| \bigvee_i (e-h_i)^+ \| = \| (e - \bigwedge_i h_i)^+ \| \leq \varepsilon'$. Consequently,

$$T_1e(f) - \varepsilon \leq \inf_{(h_i) \in H_{e,\varepsilon'}} (\bigwedge_i Th_i)(f) \leq \hat{T}(e,f).$$

The inequality $T_1e(f) \leq \hat{T}(e,f)$ being obvious for $(e,f) \in E \times (F \setminus F_+)$, T_1 is $\hat{T}$-dominated.

Conversely, if $T_1 : E \to F'$ is a $\hat{T}$-dominated operator, then we deduce from 3.22

$T_1h(f) \leq \hat{T}(h,f) = Th(f)$,

$T_1(-h)(f) \leq \hat{T}(-h,f) = T(-h)(f)$ for each $f \in F_+$ and

$Th = -T(-h) \leq T_1h \leq Th$ for each $h \in H$,

which yields $T_1|_H = T$. Moreover, for each $e \in E_+$, $f \in F_+$ we obtain $T_1(-e)(f) \leq \hat{T}(-e,f) \leq \hat{T}(0,f) = 0$, since $e' \to \hat{T}(e',f)$ is increasing. This implies that $T_1(-e) \leq 0$, which proves that T_1 is a positive operator. ∎

Although the central problem of determining the regularization $\tilde{p}^{\otimes}$ of a bisublinear functional $p : E \times F \to \mathbb{R}_\infty$ has been considerably simplified under the assumptions of Theorems 3.19 and 3.20, a complete solution has not been given until now. Promising as it may be, it is not true, however, that l.s.c. bisublinear functionals p satisfying the assumptions of 3.19 and 3.20 are regularized at least on $E \otimes F_+$, i.e.

$$\tilde{p}^{\otimes}(e \otimes f) = p^{\otimes}(e \otimes f) = p(e,f) \quad \text{for all } e \in E,\ f \in F_+.$$

This is demonstrated in the following

<u>3.24 Counterexample</u>: Let E denote an infinite dimensional reflexive Banach lattice not isomorphic to an L^p-space for any $p \in]1,\infty[$. A well-

known theorem of Tzafriri (cf.[69]) then states that there exists a closed vector sublattice H of E which is not the range of any positive projection defined on E. Hence the identity mapping $T : H \to H$ cannot be extended linearly to a positive operator $T_1 : E \to H$. On the other hand, the identity mapping $T_o : E \to E$ is a positive extension of T with range E. Identifying the reflexive spaces E and H with their respective biduals E", H" and setting F := H', G := E, we see that the bisublinear functional $\hat{T} : E \times F \to \mathbb{R}_\infty$ defined in 3.21 has all the properties mentioned in Lemma 3.22. In particular, $\hat{T}$ is subbilinear on $E \times F_+$, satisfies all assumptions of 3.20 and is l.s.c. on $E \times F$.
But the function $\tilde{\hat{T}}^{\otimes}$ attains the value $-\infty$ at 0, since, by Lemma 3.23, there is no $\hat{T}$-dominated operator $T_1 : E \to H$.

Discouraging as this counterexample may be at first sight, it gives us some useful hints. In fact, the exclusion of the classical Banach lattices rises the question, whether positive results are available at least for L^p-spaces. In the next two sections we shall see that this approach proves rather successful.
Moreover, we are now prepared to show that the usual generalization of the lower semi-continuity to sublinear mappings into embedding cones is not adequate to extend Theorem 2.9 to arbitrary Banach space valued operators (cf.[58],[59]).

<u>3.25 Corollary of 3.24</u>: Given E, F, H as in 3.24 let again $T : H \to H$ denote the identity mapping. By Theorem 3.14 there exists a sublinear mapping Φ_T form E into the sup-completion F'_s of F' such that

$$\hat{T}(e,f) = \Phi_T(f) \quad \text{for each } (e,f) \in E \times F_+.$$

Φ_T is l.s.c. at 0 in the following sense:
For each $\varepsilon > 0$, there exists $\delta > 0$ such that $\|\Phi_T(e)^-\| \leq \varepsilon$ whenever $e \in E$ satisfies $\|e\| \leq \delta$.

Indeed, since $\hat{T}$ is l.s.c. at (0,0), we can choose $\eta > 0$ such that $\hat{T}(e,f) \geq -\varepsilon$ for all $e \in E$, $f \in F$, satisfying $\|e\| \leq \eta$, $\|f\| \leq \eta$. For each $e \in E$, $\|e\| \leq \delta := \eta^2$, we hence obtain

$$\|\Phi_T(e)^-\| = \|-\Phi_T(e)^-\| = -\inf\{\Phi_T e(f) : f \in F_+, \|f\| \leq 1\}$$

$$= \sup\{-\hat{T}(e,f) : f \in F_+, \|f\| \leq 1\}$$

$$= \sup\{-T(\tfrac{1}{\eta}e, \eta f) : f \in F_+, \|f\| \leq 1\} \leq \varepsilon.$$

On the other hand, by Example 3.24, Φ_T-dominated, linear extensions $T_1 : E \to H$ of T cannot exist. ∎

Finally, we shall conclude this section with a description of $T^{\otimes}$ similar to that in Lemma 3.22, which will be used later.

<u>3.26 Theorem</u>: Let T be a positive continuous operator from a linear subspace H of some normed vector lattice E into a Dedekind complete Banach lattice F. If T has a positive continuous linear extension $T_o : E \to F'$, then the equality

$$\tilde{\hat{T}}^{\otimes}(e \otimes f) = \sup_{\varepsilon > 0} \inf_{(h_i, f_i) \in H^{\otimes}_{e,f,\varepsilon}} \sum_i Th_i(f_i)$$

holds for each $(e,f) \in E \times F_+$, where $H^{\otimes}_{e,f,\varepsilon}$ denotes the set of finite sequences $(h_i, f_i)_{i \in I}$ in $H \times F_+$ such that $(e - h_i, f_i) \in O^+_{\varepsilon,f}$. Moreover, the regularization $\Phi^{\cap}_T$ of the sublinear mapping $\Phi_T : E \to F'_s$ associated to $\hat{T}$ by Theorem 3.14 is given by

$$\Phi^{\cap}_T(e)(f) = \tilde{\hat{T}}^{\otimes}(e \otimes f) \quad \text{for all } e \in E,\ f \in F_+.$$

<u>Proof</u>: Since the set $\{Th : h \in H,\ h \geq e\}$ is bounded from below by $T_o e$ in the sup-completion F'_s of F', we can define $q_T : E \to F'_s$ by setting

$$q_T(e) = \begin{cases} \inf\{Th : h \in H,\ h \geq e\}, & \text{if } \{h \in H : h \geq e\} \neq \emptyset \\ \infty & \text{else} \end{cases}$$

Then q_T is sublinear and increasing. From Corollary 3.20 we deduce

$$\tilde{q}_T^{\otimes}(e \otimes f) = \sup_{\varepsilon>0} \inf_{(e_i,f_i)\in O^+_{\varepsilon,f}} \sum_i q_T(e-e_i)(f_i) \quad \text{for all } (e,f) \in E \times F_+$$

where $\tilde{q}_T^{\otimes}$ denotes the greatest l.s.c. function dominated by the sublinear function $q_T^{\otimes} : E \otimes F \to \mathbb{R}_\infty$ associated with the bisublinear functional $(e',f') \to q_T(e')(f')$. The inequality

$$\hat{T}(e',f') \leq q_T(e')(f') \qquad ((e',f') \in E \times F_+)$$

yields $\tilde{\hat{T}}^{\otimes}(t) \leq \tilde{q}_T^{\otimes}(t)$ for all $t \in E \otimes F$. Since $q_T(-e) \leq 0$ for each $e \in E_+$ and $q_T(h) = Th$ for all $h \in H$, every q_T-dominated, continuous operator $T_1 : E \to F'$ is a positive extension of T. By Lemma 3.23 and Theorem 2.12 we hence obtain $\tilde{\hat{T}}^{\otimes}(t) \geq \tilde{q}_T^{\otimes}(t)$ for all $t \in E \otimes F$, which implies $\tilde{\hat{T}}^{\otimes} = \tilde{q}_T^{\otimes}$. Thus it remains to show that

$$\tilde{q}_T^{\otimes}(e \otimes f) = \sup_{\varepsilon>0} \inf_{(h_i,f_i)\in H^{\otimes}_{e,f,\varepsilon}} \sum_i Th_i(f_i) \quad \text{for all } e \in E,\ f \in F_+.$$

Since $q_T(h)(f') = Th(f')$ for $h \in H$, $f' \in F_+$, the inequality

$$\tilde{q}^{\otimes}(e \otimes f) \leq \sup_{\varepsilon>0} \inf_{(h_i,f_i)\in H^{\otimes}_{e,f,\varepsilon}} \sum_i Th_i(f_i)$$

is obvious.

To prove the converse inequality, suppose that $\tilde{q}^{\otimes}(e \otimes f) <$ $< \sup_{\varepsilon>0} \inf_{(h_i,f_i)\in H^{\otimes}_{e,f,\varepsilon}} \sum_i Th_i(f_i)$ and choose $\varepsilon > 0$, $\alpha \in \mathbb{R}$ such that

$$\tilde{q}^{\otimes}(e \otimes f) < \alpha < \inf_{(h_i,f_i)\in H^{\otimes}_{e,f,\varepsilon}} \sum_i Th_i(f_i).$$

Select a finite sequence $(e_i,f_i)_{i\in I} \in O^+_{f,\varepsilon}$ such that $\sum_i q_T(e-e_i) < \alpha$. By the definition of q_T, for each $i \in I$ there exists a finite family $(h_{ij})_{j\in J_i}$ in $\{h \in H : h \geq e - e_i\}$ satisfying

$$\sum_i \Big(\bigwedge_{j\in J_i} Th_{ij}\Big)(f_i) < \alpha.$$

Since F is Dedekind complete we know that

$$(\ell_1 \wedge \ell_2)(f') = \inf\{\ell_1(f_1) + \ell_2(f_2) : f_1, f_2 \in F_+, f' = f_1 + f_2, f_1 \wedge f_2 = 0\}$$

for any two order bounded linear forms $\ell_1, \ell_2 \in F'$ and each $f' \in F_+$. Thus, for each $i \in I$, we can find an orthogonal family $(f_{ij})_{j \in J_i}$ in F_+ such that $\sum_{j \in J_i} f_{ij} = f_i$ and

$$\sum_i \left(\sum_{j \in J_i} Th_{ij}(f_{ij}) \right) < \alpha.$$

Moreover, for the bisublinear functional $r : E \times F \to \mathbb{R}_\infty$ introduced on page 52 we obtain

$$r^{\otimes}\left(\sum_i \sum_{j \in J_i} (e - h_{ij}) \otimes f_{ij}\right) =$$

$$= \sup\{\sum_i \sum_{j \in J_i} T_1(e - h_{ij})(f_{ij}) : T_1 : E \to F' \text{ positive contraction}\}$$

$$\leq \sup\{\sum_i \sum_{j \in J_i} T_1 e_i(f_{ij}) : T_1 : E \to F' \text{ positive contraction}\}$$

$$= r^{\otimes}\left(\sum_i \sum_{j \in J_i} e_i \otimes f_{ij}\right) = r^{\otimes}\left(\sum_i e_i \otimes f_i\right) \leq \varepsilon.$$

Hence, if we rewrite the double family $((h_{ij}, f_{ij})_{j \in J_i})_{i \in I}$ as a finite sequence $(h_k, f_k)_{1 \leq k \leq n}$, then (h_k, f_k) is a member of $H^{\otimes}_{e,f,\varepsilon}$ and $\sum_k Th_k(f_k) < \alpha$ contradicting the inequality

$$\inf_{(h_i', f_i') \in H^{\otimes}_{e,f,\varepsilon}} \sum_i Th_i'(f_i') > \alpha.$$

Consequently, by Theorem 2.12,

$$\sup_{\varepsilon > 0} \inf_{(h_i, f_i) \in H^{\otimes}_{e,f,\varepsilon}} \sum_i Th_i(f_i) =$$

$$= \sup\{T_1 e(f) : T_1 : E \to F' \text{ continuous, linear, } \hat{T}\text{-dominated}\}$$

$$= \sup\{T_1 e(f) : T_1 : E \to F' \text{ continuous, linear, } \Phi_T\text{-dominated}\}$$

for each $e \in E$, $f \in F_+$.

Noting that the set

$$\{T_1 e : T_1 : E \to F' \text{ continuous, linear, } \Phi_T\text{-dominated}\}$$

is upward directed by Lemma 2.11 and Theorem 2.8 we conclude

$$\widetilde{T}^{\otimes}(e \otimes f) = \sup\{T_1 e(f) : T_1 : E \to F' \text{ continuous, linear, } \Phi_T\text{-dominated}\}$$

$$= \Phi_T^{\cap}(e)(f). \quad \blacksquare$$

Final remarks to section 3:

a) Superficially considered the reader might feel that the statements of Lemma 3.3 and Remark 3.10, ii are almost contradictory. The reason is that we are accustomed to an (often undeliberate) identification of the tensor product $H \otimes F$ with a linear subspace of $E \otimes F$, when H is a linear subspace of E. Since a more detailed investigation of the relationship between $H \otimes F$ and $E \otimes F$ is crucial for understanding the problems of operator extension, the following short summary might be useful:

Given vector spaces E,F and a linear subspace $H \subset E$ consider the subspace $G \subset E \otimes F$ of all tensors $t \in E \otimes F$ that have a representation $t = \sum_i h_i \otimes f_i$ for some $h_i \in H$, $f_i \in F$. Then there exists an isomorphism $V : H \otimes F \to G$.

Let now $b : H \times F \to \mathbb{R}$ be a bilinear form and $p : E \times F \to \mathbb{R}$ denote a bisublinear form (or even a subbilinear form) such that

$$b(h,f) \leq p(h,f) \quad \text{for all } h \in H,\ f \in F.$$

Then, by Lemma 3.3, the inequality $b^{\otimes} \leq (p|_{H \times F})^{\otimes}$ holds on $H \otimes F$. This immediately yields

$$b^{\otimes} \circ V^{-1} \leq (p|_{H \times F})^{\otimes} \circ V^{-1} \quad \text{on} \quad G.$$

Thus, if the equality

$$(p|_{H \times F})^{\otimes} \circ V^{-1} = p^{\otimes}|_G$$

were true, where $p^{\otimes}$ is formed on $E \otimes F$, then as an immediate consequence of the Hahn-Banach theorem we were able to extend b bilinearly under domination by p to $E \times F$ contradicting the counterexample established in 3.10, ii. The respective transitions from p and $p|_{H \times F}$ to $p^{\otimes}$ and $(p|_{H \otimes F})^{\otimes}$ are therefore incompatible with the isomorphism V. The reason for this phenomenon is that for $t \in G$ $p^{\otimes}(t)$ is formed by using all tensor product representations $t = \sum_i e_i \otimes f_i$, $e_i \in E$, $f_i \in F$ not only those for which $e_i \in H$.

b) In view of Theorem 3.19 there is no need to restrict our attention only to the extension of positive operators. Hopefully, that the approach towards solving extension problems demonstrated in the next two sections will also contribute to the solution of non-positive extension problems we deferred the restriction to positive operators as far as possible.

4. Extension of L^1-valued positive operators

As a first test of the efficiency of the results obtained in the preceding sections we start with the solution of the existence problem for norm-preserving positive extensions of operators with values in some AL-space. For merely positive extensions (without restrictions on the norm) we not only provide a surprisingly simple condition for the existence of extensions but also determine the linear subspace up to which the extensions are unique.

4.1 Definition: Let (E,G) denote a pair of Banach lattices possessing the following properties:

(4.1.1) If $J : G \to G''$ denotes the natural embedding of G into its topological bidual, then there exists a positive contractive projection from G'' onto $J(G)$.

(4.1.2) In the topological dual $F := G'$ there exists an upward directed net $(F_j)_{j\in J}$ of finite dimensional vector sublattices such that $\bigcup_{j\in J} F_j$ is dense in F.

(4.1.3) For each finite family $(e_i, f_i)_{i\in I}$ in $E \times F_+$ satisfying $e_i \neq 0$, $f_i \neq 0$ and $f_i \wedge f_j = 0$ for all $i,j \in I$, $j \neq i$, there is a family $(\lambda_i)_{i\in I}$ of strictly positive real numbers such that

$$\|\bigvee_{i\in I} \lambda_i e_i^+\| \, \|\sum_{i\in I} \frac{1}{\lambda_i} f_i\| \leq r^{\otimes}(\sum_i e_i \otimes f_i),$$

where $r : E \times F \to \mathbb{R}_\infty$ denotes the bisublinear functional introduced before Corollary 3.20.

Then (E,G) will be called an adapted pair (of Banach lattices).

4.2 Remarks:

a) It is well-known that condition (4.1.1) holds for Dedekind complete AM-spaces as well as for all KB-spaces in Vulikh's terminology (see [71], p. 188). In particular, all L^p-spaces, $1 \leq p < \infty$ and also $L^\infty(\mu)$ for a localizable measure μ admit a positive projection $P : G'' \to J(G)$.

b) Condition (4.1.3) is the main restriction on the pair (E,G). It states in fact that

$$r^{\otimes}(\sum_i e_i \otimes f_i) = \inf\{\| \bigvee_{i\in I} \lambda_i e_i^+\| \cdot \|\sum_i \frac{1}{\lambda_i} f_i\| : (\lambda_i)_{i\in I} \in]0,\infty[^I\}.$$

To show this we claim that, under the assumptions of (4.1.3), for any pair (E,G) of Banach lattices the inequality

$$(+) \quad r^{\otimes}(\sum_i e_i \otimes f_i) \leq \inf\{\| \bigvee_i \lambda_i e_i^+\| \cdot \|\sum_i \frac{1}{\lambda_i} f_i\| : (\lambda_i)_{i\in I} \in]0,\infty[^I\}$$

holds. Since $r^{\otimes}(\sum_i e_i \otimes f_i)$ is the supremum of all values $\sum_i Te_i(f_i)$ where T ranges over all positive contractive operators from E into $F' = G''$, this inequality follows from the estimate

$$\sum_i Te_i(f_i) = \sum_i T(\lambda_i e_i)(\frac{1}{\lambda_i} f_i) \leq (\bigvee_i T(\lambda_i e_i)^+)(\sum_i \frac{1}{\lambda_i} f_i) \leq T(\bigvee_i \lambda_i e_i^+)(\sum_i \frac{1}{\lambda_i} f_i)$$

$$\leq \| \bigvee_i \lambda_i e_i^+\| \cdot \|\sum_i \frac{1}{\lambda_i} f_i\|$$

valid for each family $(\lambda_i)_{i\in I} \in]0,\infty[^I$ and every positive contraction from E into G".

Thus, condition (4.1.3) indicates that the pair (E,G) is minimal with respect to the inequality (+).

The following theorem presents a selection of adapted pairs of Banach lattices. In the next section we shall add further important examples.

<u>4.3 Theorem</u>: For every Banach lattice E and every AL-space G (E,G) is an adapted pair.

<u>Proof</u>: Since every AL-space is a KB-space (see [66], Ch. II, 8.3), condition (4.1.1) holds. In order to show (4.1.2) note that $F := G'$ is an AM-space with unit. Hence there exists a compact Stonian space X, such that F is isometrically isomorphic to $\mathcal{C}(X)$. Let $\mathfrak{Z}$ denote the system of all finite partitions of X into closed-open subsets. If, for each $Z \in \mathfrak{Z}$, $\tilde{F}_Z$ is the linear space of all functions $f \in \mathcal{C}(X)$ constant on each set of the partition Z and if F_Z is the corresponding linear subspace of F, then $(F_Z)_{Z\in\mathfrak{Z}}$ is clearly an increasing net of vector sublattices of F with dense union.

Finally, to prove condition (4.1.3) let $(e_i, f_i)_{i\in I}$ be a finite family in $E \times (F_+ \setminus \{0\})$ such that $f_i \wedge f_j = 0$ for all $i,j \in I$, $i \neq j$. Consider a family $(f'_i)_{i\in I}$ of positive linear forms on F satisfying

i) $\|f'_i\| = 1$ and $f'_i(f_i) = \|f_i\|$ for all $i \in I$,

ii) $f'_i \wedge f'_j = 0$ for all $i,j \in I$, $i \neq j$.

Choose a positive linear form ℓ on E such that $\|\ell\| \leq 1$ and $\ell(\bigvee_i \|f_i\| e_i^+) = \|\bigvee_i \|f_i\| e_i^+\|$.

Then there exists an orthogonal family $(\ell_i)_{i\in I}$ of positive linear forms on E with sum $\sum_{i\in i} \ell_i = \ell$ satisfying the equality

$$\sum_i \ell_i(\|f_i\| e_i) = \ell(\bigvee_i \|f_i\| e_i^+)$$

(concerning the existence of such a family see, e.g., [66], Ch. II., Cor. of Prop. 4.10). The positive operator $T : E \to F'$, defined by

$$Te = \sum_i \ell_i(e) \cdot f'_i \qquad (e \in E),$$

is contractive. Indeed, since the norm of F' is additive on F'_+, we obtain

$$\|Te\| = \sum_i \ell_i(e)\|f_i'\| = \sum_i \ell_i(e) = \ell(e) \le 1$$

for each $e \in E_+$. Consequently,

$$\|\bigvee_i \|f_i\| e_i^+\| = \ell(\bigvee_i \|f_i\| e_i^+) = \sum_i \ell_i(\|f_i\| e_i) = \sum_i \ell_i(e)\|f_i\| =$$

$$= \sum_i Te_i(f_i) \le r^{\otimes}(\sum_i e_i \otimes f_i),$$

which implies

$$\|\bigvee_i \lambda_i e_i^+\| \, \|\sum_i \frac{1}{\lambda_i} f_i\| \le r^{\otimes}(\sum_i e_i \otimes f_i) \quad \text{for} \quad \lambda_i := \|f_i\|,$$

since F is an AM-space. ■

The following theorems show that, once we have proved the adapted pair property, positive and norm-preserving extensions of operators are no longer hard to find.

4.4 Theorem: Given an adapted pair (E,G) of Banach lattices and a linear subspace H of E, let $T : H \to G$ be a positive continuous operator. For each $M \in \mathbb{R}_+$ such that $\|T\| \le M$, the following are equivalent:

i) $\|\bigvee_i (Th_i)^+\| \le M\|\bigvee_i h_i^+\|$ for every finite family (h_i) in H.

ii) There exists a positive extension $T_o : E \to G$ of T such that $\|T_o\| \le M$.

iii) For each $e \in E$ there is a positive, continuous extension $T_e : H + \mathbb{R}e \to G$ of T with norm $\|T_e\| \le M$.

Proof: (ii) ⇒ (iii) is obvious.

(iii) ⇒ (i): Given a finite family (h_i) in H let $e := \bigvee_i h_i^+$ and choose a positive continuous extension $T_e : H + \mathbb{R}e \to G$ of T such that $\|T_e\| \le M$. Then (i) results from the estimate

$$\|\bigvee_i (Th_i)^+\| = \|\bigvee_i (T_e h_i)^+\| \le \|T_e(\bigvee_i h_i^+)\| \le M \cdot \|\bigvee_i h_i^+\|.$$

(i) ⇒ (ii): If $(F_j)_{j\in J}$ is an increasing net of finite dimensional vector sublattices of $F := G'$ with dense union F_o, let $(h_i, f_i)_{i\in I}$ be a finite family in $H \times F_o$. Then we can find an index $j \in J$ such that $f_i \in F_j$ for all $i \in I$. For a positive orthogonal basis $(u_k)_{k\in K}$ of F_j let $(\lambda_{ik})_{k\in K} \in \mathbb{R}^K$ be the family of coordinates of f_i, i.e. $\sum_k \lambda_{ik} u_k = f_i$, for each $i \in I$. Note that the elements $\bar{h}_k := \sum_i \lambda_{ik} h_i$ are still in H for each $k \in K$, and that

$$t := \sum_i h_i \otimes f_i = \sum_k (\sum_i \lambda_{ik} u_k) \otimes u_k = \sum_k \bar{h}_k \otimes u_k .$$

If $K' := \{k \in K : \bar{h}_k^+ \neq 0\}$, we deduce from condition (4.1.3) that there is a family $(\lambda_k)_{k\in K'}$ in $]0,\infty[$ such that

$$\|\bigvee_{k\in K'} \lambda_k \bar{h}_k^+\| \cdot \|\sum_{k\in K'} \frac{1}{\lambda_k} u_k\| \leq r^{\otimes}(\sum_{k\in K'} \bar{h}_k \otimes u_k) = r^{\otimes}(\sum_{k\in K} \bar{h}_k \otimes u_k) = r^{\otimes}(t)$$

using Lemma 3.18. Therefore

$$T^{\otimes}(t) = \sum_{k\in K\setminus K'} u_k(T\bar{h}_k) + \sum_{k\in K'} (\frac{1}{\lambda_k} u_k)(T(\lambda_k \bar{h}_k)) \leq (\sum_{k\in K'} \frac{u_k}{\lambda_k})(\bigvee_{k\in K'} T(\lambda_k \bar{h}_k)^+)$$

$$\leq M \cdot \|\sum_{k\in K'} \frac{u_k}{\lambda_k}\| \cdot \|\bigvee_{k\in K'} \lambda_k \bar{h}_k^+\| \leq M \cdot r^{\otimes}(t),$$

which implies that $T^{\otimes} \leq M \cdot r^{\otimes}$ on $H \otimes F_o$. By the Hahn-Banach theorem there exists a $M\cdot r$-dominated linear extension $T_1 : E \to F' = F_o'$ of T. The remarks preceding Corollary 3.20 show that T_1 is in fact positive, continuous and $\|T_1\| \leq M$.

Finally, if $J : G \to G'' = F'$ denotes the natural imbedding and $P : G'' \to J(G)$ is a positive contractive projection onto $J(G)$, then $T_o := J^{-1} \circ P \circ T_1 : E \to G$ is a positive extension of T with norm $\|T_o\| \leq M$. ∎

We shall defer the formulation of some important consequences of Theorem 4.4 to the next section when we have supplied a rich choice of examples for adapted pairs of Banach lattices. Here we shall only formulate a result of Lotz (see [49]) immediately following from Thm.4.4:

4.5 Corollary: Every continuous positive operator T from a vector sublattice H of a Banach lattice E into an AL-space G has a norm-preserving positive extension to E.

4.6 Remarks:

a) At first sight, it may seem that, for the proof of Theorem 4.4, the whole theoretical background developed in section 3 is necessary. A more detailed investigation shows, however, that we only need the introduction of the bisublinear functional r on page 52 and Theorem 4.3 in order to prove Theorem 4.4.

b) Condition (iii) of Theorem 4.4 is closely related to a conjecture of Nachbin (see [55]) on norm-preserving extensions in Banach spaces. For a linear subspace H of a Banach space E and a continuous operator T from H into some Banach space G Nachbin asked whether the following condition is sufficient for the existence of a norm-preserving linear extension of T to E:

For each $e \in E$ there is a norm-preserving extension $T_e : H + \mathbb{R}e \to G$ of T.

Lindenstrauß has shown (see [46]) that this condition does not imply the existence of norm-preserving linear extension of T to E, in general, by an instructive counterexample. On the other hand, Theorem 4.4 reveals that Nachbin's condition is sufficient for the existence of norm-preserving extensions of positive operators between adapted Banach lattices.

The following theorem provides a surprisingly simple solution of the existence problem for positive linear extensions between adapted Banach lattices.

4.7 Theorem: Let (E,G) be an adapted pair of Banach lattices. If H is a linear subspace of E and $T : H \to G$ is a positive operator, the following conditions are equivalent:

i) There exists a real number $M \geq 0$ such that

$\| \bigvee_i (Th_i)^+ \| \leq M \| \bigvee_i h_i^+ \|$ for every finite family (h_i) in H.

ii) There exists a positive extension $T_o : E \to G$ of T.

iii) If, for each $a \in E_+$, $E_a := \{e \in E : \exists\, \lambda > 0 : |e| \leq \lambda a\}$ denotes the vector lattice ideal generated by a, then there exists a positive linear extension of $T|_{H \cap E_a}$ to E_a.

iv) For every subset A of H bounded from above in E T(A) is bounded from above in G.

Proof: The implications (ii) ⇒ (iii), (ii) ⇒ (iv) are evident, (i) ⇒ (ii) is an immediate consequence of Theorem 4.4.

(iii) ⇒ (i): Suppose (iii) were true, but condition (i) failed. For each $n \in \mathbb{N}$ we then could find a finite family $(h_i^{(n)})_{i \in I_n}$ in H such that

$$\| \bigvee_{i \in I_n} (Th_i^{(n)})^+ \| > n \cdot 2^n \| \bigvee_{i \in I_n} h_i^{(n)+} \| .$$

Setting $k_n := \frac{1}{2^n} \dfrac{\bigvee_i h_i^{(n)+}}{\| \bigvee_i h_i^{(n)+} \|}$ for each $n \in \mathbb{N}$, the series $\sum_{n=1} k_n$ is absolutely convergent in E. If $a := \sum_{n=1}^{\infty} k_n$, choose a positive operator $T_a : E_a \to G$ extending $T|_{H \cap E_a}$. Note that $k_n \in E_a$ for each $n \in \mathbb{N}$ and that

$$2^n \| T_a k_n \| = \frac{\| T_a (\bigvee_{i \in I_n} h_i^{(n)+}) \|}{\| \bigvee_{i \in I_n} h_i^{(n)+} \|} \geq \frac{\| \bigvee_i T_a (h_i^{(n)})^+ \|}{\| \bigvee_i h_i^{(n)+} \|} \geq 2^n \cdot n .$$

Since $k_n \leq a$, we conclude

$\| T_a(a) \| \geq n$ for each $n \in \mathbb{N}$, which is absurd.

(iv) ⇒ (iii): Given $a \in E_+$ let $H_a := H \cap E_a$. By condition (iv) there exists an element $b \in G$ such that $Th \leq b$ for all $h \in H_a$, $h \leq a$. Hence, if we define

$$\varphi(e) = \inf\{\lambda > 0 : \lambda a \geq e\} \qquad (e \in E),$$

then $Th \leq \varphi(h)\cdot b$. Moreover, $e \to \varphi(e)\cdot b$ is a sublinear mapping from E_a into G. Since G is Dedekind complete by condition 4.1.1, the theorem of Hahn-Banach shows that there exists an operator $T_a : E_a \to G$ such that $T_a h = Th$ for all $h \in H_a$ and $T_a e \leq \varphi(e)\cdot b$ for all $e \in E$. From the inequality $T_a(-e) \leq \varphi(-e)\cdot b = 0$, valid for each $e \in E_+$, we finally conclude that T_a is positive. ∎

The following slight modification of Counterexample 3.24 demonstrates the statement that Theorem 4.7 is not true for arbitrary Banach lattices E, G:

<u>4.8 Counterexample</u>: Let E denote a Banach lattice with order continuous norm which is not isomorphic (as a topological vector lattice) to any L^p-space, $p \in [1,\infty]$, or to c_o, the space of all real sequences vanishing at infinity. Then there exists a closed vector sublattice H of E, which is not the range of a positive projection $P : E \to H$ (cf.[69]). If we set $G := H$ the identity mapping $T : H \to H$ satisfies condition (iv) of Theorem 4.7 (and also the conditions (i), (iii). Indeed, if $A \subset H$ is bounded from above in E, sup A exists in E by the Dedekind completeness of E. Furthermore, $\sup A = \lim_{i \in I} k_i$, where $(k_i)_{i \in I}$ denotes the increasing net of all suprema of finite subsets of A. The vector sublattice $H = G$ being closed we obtain $\lim_{i \in I} k_i \in G$. Thus $A = T(A)$ is bounded from above in G. But there is no positive linear extension $T_o : E \to G$ of T, for T_o would be a positive projection.

<u>4.9 Remark</u>: It is easy to generalize the statements of 4.4 and 4.7 to regular operators (i.e. differences of positive operators). For example, Theorem 4.4 allows the following obvious modification concerning the extension of regular operators between an adapted pair of Banach lattices (E,G):

If $T : H \to G$ is regular and $M \geq 0$ is a non-negative constant, then the following are equivalent:

i') $\| \bigvee_i |Th_i| \| \leq M \| \bigvee_i |h_i| \|$ for every finite family (h_i) in H.

ii') There exists a regular linear extension $T_o : E \to G$ of T such that $\| |T_o| \| \leq M$.

iii') For each $e \in E$ there is a regular, continuous, linear extension $T_e : H + \mathbb{R}e \to G$ of T such that $\| |T_e| \| \leq M$.

The corresponding modifications of Theorem 4.7 are evident (e.g. condition (iv) changes to (iv'): T(A) is order bounded in G whenever $A \subset H$ is order bounded in E). Note that, for AL-spaces E, every continuous linear operator $T : E \to G$ is regular, provided that G satisfies condition 4.1.1. The following theorem of M. Levy (cf. [44]) is thus an immediate consequence of the modification of Theorem 4.7 just mentioned:

If E and G are AL-spaces and T is a regular operator from a linear subspace H of E into G then there exists a continuous linear extension $T : E \to G$, provided that

$$\| \bigvee_i |Th_i| \| \leq M \| \bigvee_i |h_i| \| \quad \text{for every finite family } (h_i) \text{ in } H$$

and some $M \geq 0$ (independent of (h_i)).

Somewhat more instructive is the equivalent condition that T(A) should be order bounded in G whenever $A \subset H$ is order bounded in E.

Restricting our attention to positive operators the problems of the last two sections can be satisfactorily solved for adapted pairs of Banach lattices:

4.10 Theorem: Let E,G be Banach lattices such that the topological dual $F := G'$ of G satisfies condition (4.1.3). Furthermore, let $p : E \times F \to \mathbb{R}_\infty$ be a l.s.c. bisublinear functional possessing the following properties:

i) $\{f \in F : p(e,f) < \infty\} \subset F_+$ for each $e \in E \setminus \{0\}$,

ii) $f \to p(e,f)$ is additive on F_+ for each $e \in E$,

iii) $e \to p(e,f)$ is increasing for each $f \in F_+$.

Then $\widetilde{p}^{\otimes}(e \otimes f) = p(e,f)$ for all $e \in E$, $f \in F_+$.

Proof: Let us first prove the lower semi-continuity of $p^{\otimes}$ in O. If $\alpha < 0$, there exists $\varepsilon > 0$ such that $p(e,f) \geq \alpha$ for all $(e,f) \in E \times F$ satisfying $\|e\|\|f\| \leq \varepsilon$, p being bisublinear and l.s.c. at O.
Let $(e_i,f_i)_{i\in I} \in O_\varepsilon^+$, where O_ε^+ denotes the set introduced in Corollary 3.20. If $I' := \{i \in I : e_i^+ \neq 0 \text{ and } f_i \neq 0\}$ there exists a finite family $(\lambda_i)_{i\in I'} \in]0,\infty[^{I'}$ such that

$$\|\bigvee_{i\in I'} \lambda_i e_i^+\| \|\sum_{i\in I'} \frac{1}{\lambda_i} f_i\| \leq r^{\otimes}(\sum_{i\in I'} e_i \otimes f_i) = r^{\otimes}(\sum_{i\in I} e_i \otimes f_i) \leq \varepsilon$$

using condition (4.1.3) and Lemma 3.18. The functional $e \to p(e,f)$ being isotone for each $f \in F_+$ we conclude

$$\sum_{i\in I} p(-e_i,f_i) \geq \sum_{i\in I'} p(-e_i,f_i) = \sum_{i\in I'} p(-\lambda_i e_i, \frac{1}{\lambda_i} f_i) \geq$$

$$\geq \sum_{j\in I'} p(-\bigvee_{i\in I'} \lambda_i e_i, \frac{1}{\lambda_j} f_j) \geq p(-\bigvee_{i\in I'} \lambda_i e_i^+, \sum_{j\in I'} \frac{1}{\lambda_j} f_j) \geq \alpha .$$

Therefore, $\widetilde{p}^{\otimes}(0) = \sup_{\varepsilon'>0} \inf_{(e_i,f_i)\in O_\varepsilon^+} \sum_i p(-e_i,f_i) \geq \alpha$ by Theorem 3.20, which shows that $p^{\otimes}$ is l.s.c. at O.
Applying once more Theorem 3.20 and observing condition 4.1.3, similarly yields

$$p(e,f) = \sup_{\varepsilon>0} \inf_{(e_i,f_i)\in O_{\varepsilon,f}^+} \sum_i p(e - e_i,f_i) = \widetilde{p}^{\otimes}(e \otimes f) \text{ for all } e \in E,\ f \in F_+ . \blacksquare$$

As an application of Theorem 4.10 we shall now solve the uniqueness problem for positive linear extensions of operators between adapted Banach lattices E,G, when G is a KB-space.

Consider a linear subspace H of E and a positive continuous operator $T : H \to G$ possessing a positive linear extension $T_o : E \to G$. We wish to characterize all those elements $e \in E$, satisfying the coincidence condition

$$T_1 e = T_o e$$

for all positive linear extensions $T_1 : E \to G$ of T. More generally, we wish to determine the set

$$\{T_1 e : T_1 : E \to G \ \text{ positive linear extension of } T\}.$$

Setting $F := G'$ and imbedding G into its bidual $G'' = F'$ makes T an operator from H into F'. The functional $\hat{T} : E \times F \to \mathbb{R}_\infty$, defined by

$$\hat{T}(e,f) = \begin{cases} \sup\limits_{\varepsilon>0} \inf\limits_{(h_i) \in H_{e,\varepsilon}} f(\bigwedge\limits_i Th_i) & \text{for } e \in E,\ f \in F_+ \\ 0 & \text{if } e = 0,\ f \in F \setminus F_+, \\ \infty & \text{whenever } e \in E \setminus \{0\},\ f \in F \setminus F_+ \end{cases}$$

satisfies the assumptions of Theorem 4.10 by Lemma 3.22.

With these notations we obtain

<u>4.11 Theorem</u>: Given $e \in E$, $f \in F_+$ and $\alpha \in \mathbb{R}$ such that

$$-\hat{T}(-e,f) < \alpha < \hat{T}(e,f),$$

there exists a positive extension $T_1 : E \to G$ of T satisfying $f(T_1 e) = \alpha$. If $e \in E$ is fixed, all positive linear extensions T_1 of T coincide at e provided that $-\hat{T}(-e,f) = \hat{T}(e,f)$ for all $f \in F$.

<u>Proof</u>: From Theorem 4.10 we deduce that $\tilde{\hat{T}}^{\otimes}(e \otimes f) = \hat{T}(e,f) > \alpha$ and $\tilde{\hat{T}}^{\otimes}(-e \otimes f) = \hat{T}(-e,f) > -\alpha$. Hence, by Corollary 2.9, there exists a $\hat{T}$-dominated operator $S : E \to F'$ such that $Se(f) = \alpha$. Let $J : G \to G''$ de-

note the natural imbedding. Since G is a KB-space, J(G) is a band in $G'' = F'$. Setting $R := J^{-1} \circ P \circ S$, where $P : G'' \to J(G)$ denotes the band projection, we obtain $Rh = J^{-1}(Sh) = Th$ for all $h \in H$. Hence, if $f(Re) = \alpha$, the proof is complete with $T_1 := R$.
If $f(Re) < \alpha$ we put $x := (Se - (P \circ S)(e))^+$.
Note that $x \neq 0$, since $(Se - (P \circ S)(e))(f) = \alpha - f(Re) > 0$.
There exists a positive linear form φ on G" such that $\varphi(x) = 1$ and $\varphi(y) = 0$ for all $y \in G''$ satisfying $x \wedge |y| = 0$. Selecting an element $g_o \in G_+$ such that $f(g_o) = 1$ we set

$$T_1 e' := Re' + (\alpha - f(Re))\varphi(Se') \cdot g_o \quad \text{for all } e' \in E.$$

Then $T_1 h = Rh = Th$ for each $h \in H$, since $Sh = Th \in J(G)$ and $x \wedge |y| = 0$ for all $y \in J(G)$. Finally, we check

$$f(T_1 e) = f(Re) + (\alpha - f(Re))\varphi(Se) = f(Re) + (\alpha - f(Re))\varphi((Se - P \circ Se)^+) = \alpha.$$

If $f(Re) > \alpha$, we define T_1 similarly replacing φ by a positive linear form ψ on G" satisfying

$$\psi((Se - P \circ Se)^-) = 1,$$

$\psi(y) = 0$ for all $y \in G''$ such that $(Se - P \circ Se)^- \wedge |y| = 0$.

The second part of the assertion is an immediate consequence of 3.21.■

As a further consequence of Theorem 4.10 we obtain the following result formulated for σ-finite measure spaces only.

<u>4.12 Corollary of 4.10</u>: Let $G = L^1(\mu)$ for a σ-finite measure space $(\Omega, \mathfrak{A}, \mu)$ $F := G'$ and E an arbitrary Banach lattice. Furthermore, let $p : E \times F \to \mathbb{R}_\infty$ be a l.s.c. bisublinear functional satisfying the following conditions:

i) $\{f \in F : p(e,f) < \infty\} \subset F_+$ for all $e \in E \setminus \{0\}$,

ii) $f \to p(e,f)$ is additive on F_+ for each $e \in E$,

iii) $e \to p(e,f)$ is isotone for all $f \in F_+$,

iv) $\lim_{i\in I} p(e,f_i) = p(e,f)$ for each $f \in F$ and every increasing net $(f_i)_{i\in I}$ in E such that $\sup_{i\in I} f_i = f$.

For each pair $(e,f) \in E \times F_+$ and every choice of $\alpha \in \mathbb{R}$ such that $-p(-e,f) < \alpha < p(e,f)$ there exists a p-dominated operator $T_1 : E \to G$ satisfying $f(T_1 e) = \alpha$.

Proof: Let Θ_p denote the set of all p-dominated operators $T : E \to F' = G''$. By Theorem 4.10 we know that

$$p(e,f) = \sup\{Te(f) : T \in \Theta_p\} \quad \text{for each } (e,f) \in E \times F_+ .$$

Hence, if F'_s denotes the sup-completion of F' (see Lemma 3.13) and if $p_s : E \to F'_s$ is the sublinear mapping defined by

$$p_s(e)(f) = p(e,f) \qquad ((e,f) \in E \times F_+)$$

according to Lemma 3.14, the regularization $p_s^{\cap}$ of p_s coincides with p_s. Indeed, $\{k(e) : k \in K_{p_s,e}\}$ being upward directed by Lemma 2.11 the same must be true for $\{Te : T \in \Theta_p\}$ by Theorem 2.8. Consequently,

$p_s^{\cap}(e)(f) = \sup\{Te(f) : T \in \Theta_p\} = p(e,f) = p_s(e)(f)$ for each $f \in F_+$,

which yields $p_s^{\cap}(e) = p_s(e)$.

Identifying $L^1(\mu)' = G'$ with $L^\infty(\mu)$ we define the μ-absolutely continuous (in general non-positive) measure

$$\nu(A) := p(e,1_A) \qquad (A \in \mathcal{A})$$

on $(\Omega, \mathcal{A})$, where 1_A denotes the indicator function of A. The Radon-Nikodym derivative $p_1(e)$ of ν is a member of the tight imbedding cone C_1 of $L^1(\mu)$ introduced in Example 1.2,d and

$$\int p_1(e) 1_A d\mu = p(e,1_A) \text{ for all } A \in \mathcal{A}.$$

The functional $f \to p(e,f)$ being additive on F_+ we deduce that

$$(4.12.1) \quad \int p_1(e) f \, d\mu = p(e,f) = p_s(e)(f) \quad \text{for all } f \in F_+$$

using the fact that each element $f \in L^\infty(\mu)_+ = F_+$ is the supremum of a suitable increasing sequence of simple functions. In particular,

$p_1 : E \to C_1$ is sublinear.

In order to show that p_1 is regularized, let P denotethe band projection from G" onto the canonical image J(G) in G". For simplicity, we shall no longer distinguish between G and J(G) in the rest of the proof.

Given $e \in E$ and $k \in \mathcal{K}_{p_s,e}$ the domain of k is of the form $e + U$ for some convex zero-neighborhood U in E. G being a band in G" we obtain

$$\begin{aligned} P(k(e')^+) &= \sup\{g \in G : g \le k(e')^+\} \\ &\le \sup\{g \in G : f(g) \le p_s(e')^+(f) = \int p_1(e')^+ \cdot f d\mu \text{ for all } f \in F_+\} \\ &\le p_1(e')^+ \end{aligned}$$

for each $e' \in e + U$.

Now $p_1(e')^- \in L^1(\mu) = G$, hence $P(p_1(e')^-) = p_1(e')^-$ and

$$\begin{aligned} (-k(e')^-)(f) = (k(e') \wedge 0)(f) \le (p_s(e') \wedge 0)(f) &= \\ = (p_1(e') \wedge 0)(f) &= -p_1(e')^-(f) \end{aligned}$$

for all $f \in F_+$. Consequently, $p_1(e')^- \le k(e')^-$ which implies

$$\begin{aligned} p_1(e')^- &= P(p_1(e')^-) \le P((ke')^-) \text{ and} \\ P(k(e')) &= P(k(e')^+ - k(e')^-) \le p_1(e')^+ - p_1(e')^- = p_1(e'), \end{aligned}$$

i.e. $\{P \circ k : k \in \mathcal{K}_{p_s,e}\} \subset \{k : k \in \mathcal{K}_{p_1,e}\}$.

For each $g \in G$ satisfying $g \le p_s(e)$ we thus obtain

$$\sup_{k \in \mathcal{K}_{p_s,e}} (P(k(e)) \wedge g) = P(\sup_{k \in \mathcal{K}_{p_s,e}} (k(e) \wedge g)) = Pg = g .$$

It follows that, <u>in C_1</u>, we have the equalities

$$\sup_{k \in \mathcal{K}_{p_s,e}} P(k(e)) = \sup_{k \in \mathcal{K}_{p_s,e}} \sup_{\substack{g \in G \\ g \le p_s(e)}} (P(k(e)) \wedge g) =$$

$$= \sup_{\substack{g \in G \\ g \le p_s(e)}} \sup_{k \in \mathcal{K}_{p_s,e}} (P(k(e)) \wedge g) = \sup\{g \in G : g \le p_s(e)\} = p_1(e)$$

since, by condition (4.12.1), the sets $\{g \in G : g \le p_s(e)\}$ and

$\{g \in G : g \le p_1(e)\}$ coincide. The resulting inequality

$$p_1^{\cap}(e) = \sup_{k_1 \in K_{p_1,e}} k_1(e) \ge \sup_{k \in K_{p_s,e}} P(k(e)) = p_1(e)$$

yields $p_1(e) = p_1^{\cap}(e)$.

To complete the proof let $T_o : E \to G$ be an arbitrary p_1-dominated operator and let $e \in E$, $f \in F_+$, $\alpha \in]-p(-e,f), p(e,f)[$ be fixed. By Lemma 2.11 we can find $k_1 \in K_{p_1,e}$, $k_2 \in K_{p_1,-e}$ such that

$$f(k_1(e)) > \alpha,\ k_1(e) \ge T_o e \text{ and } f(k_2(-e)) > -\alpha,\ k_2(-e) \ge T_o(-e) .$$

Choose $\lambda \in [0,1]$ such that $f(\lambda k_1(e) + (1-\lambda)(-k_2(-e))) = \alpha$. The element $g := \lambda k_1(e) + (1-\lambda)(-k_2(-e))$ then satisfies $g \le k_1(e)$ and $-g \le k_2(-e)$. Since $T_o \in K_{p_1,e}$, Theorem 2.8 shows that there exists a p_1-dominated operator $T : E \to G$ such that $Te = g$, which implies $f(Te) = f(g) = \alpha$. ∎

<u>4.13 Remark</u>: Condition (iv) of Theorem 4.12 cannot be eliminated. To show this, consider the following example:

Let $E = G = \ell^1$ and define $s : \ell^\infty \to \mathbb{R}$ by

$$s((\xi_n)_{n \in \mathbb{N}}) = \limsup_{n \to \infty} \xi_n .$$

Then s is clearly sublinear. If $F_o \subset \ell^\infty$ denotes the linear subspace of all convergent sequences and $\rho_o : F_o \to \mathbb{R}$ is the linear form

$$\rho_o((\xi_n)_{n \in \mathbb{N}}) = \lim_{n \to \infty} \xi_n ,$$

there exists a s-dominated, linear extension ρ of ρ_o to ℓ^∞ by the Hahn-Banach theorem. ρ is positive, since

$$-\rho(x) = \rho(-x) \le s(-x) \le 0 \quad \text{for each } x \in \ell^\infty_+ .$$

Setting $F := (\ell^1)' \cong \ell^\infty$ we define the bisublinear functional $p : E \times F \to \mathbb{R}_\infty$ by the equality

$$p(e,f) = \begin{cases} \|e^+\|\cdot\rho(f) & \text{for } e\in E,\ f\in F_+ \\ 0 & \text{if } e=0,\ f\in F\setminus F_+ \\ \infty & \text{else.} \end{cases}$$

Then p is l.s.c. and obviously satisfies the condition (i) - (iii) of Theorem 4.12. For $e\in E_+\setminus\{0\}$ and $f\in F_+$ such that $\rho(f) > 0$ choose $\alpha\in]0,\|e^+\|\cdot\rho(f)[$. Then

$$-p(-e,f) = -\|(-e)^+\|\cdot\rho(f) = 0 < \alpha < \|e^+\|\rho(f) = p(e,f).$$

But there are no p-dominated operators $T_1 : \ell^1 \to \ell^1$ except the zero operator. To prove this, let $x\in\ell^1_+$ be given. From the inequality $f(T_1(-x)) \leq p(-x,f) = 0$, valid for all $f\in F_+$, we deduce that $T_1(x) \geq 0$. For $f\in F_+ = (\ell^1)'_+$ and each $n\in\mathbb{N}$ we define

$$f_n((\xi_m)_{m\in\mathbb{N}}) = f(\xi_m^{(n)})_{m\in\mathbb{N}} \qquad ((\xi_m)\in\ell^1)$$

where $\xi_m^{(n)} := \xi_m$ for $m \leq n$ and $\xi_m^{(n)} := 0$ for $m > n$.

Then

$$f(T_1x) = \sup_{n\in\mathbb{N}} f_n(T_1x) \leq \sup_{n\in\mathbb{N}} p(x,f_n) = \sup_{n\in\mathbb{N}}\|x\|\cdot\rho(f_n)$$

$$\leq \sup_{n\in\mathbb{N}}\|x\|\cdot s(f_n) = 0$$

for each $f\in F_+$ which implies that $T_1x = 0$. Since $E = E_+ - E_+$, T_1 must be the zero-operator. ■

5. Extension of positive operators in L^p-spaces

Reviewing the results of the last section it has become obvious that adapted pairs of Banach lattices provide an elaborate extension theory for positive operators. On the other hand, the applications were restricted to L^1-spaces as target space, since, until now, the only examples for adapted Banach lattices were established through Theorem 4.3.

In this section we shall show that all pairs (E,G) of Banach lattices, where E is an L^p-space and G an L^q-space, $p,q, \in [1,+\infty]$, $q \leq p$, are adapted pairs of Banach lattices.

For the proof of this central theorem we need the following three lemmas.

For a finite measure space $(\Omega, \mathcal{A}, \mu)$ and a real number $p \in [1,\infty[$ consider the vector lattice $\mathcal{L}^p := \mathcal{L}^p(\mu)$ of all p^{th}-power μ-integrable, real-valued functions on Ω introduced in Example 1.2,d.

If $f,g \in \mathcal{L}^p_+$, let $f_g \in \mathcal{L}^p_+$ be given by

$$f_g(x) := \begin{cases} f(x) & \text{whenever } f(x) \geq g(x) \\ 0 & \text{else} \end{cases} \qquad (x \in \Omega).$$

With these notations we can state

5.1 Lemma: If the functions $f_o, g_o \in \mathcal{L}^p_+$ satisfy the condition

$$\mu(\{x \in \Omega : f_o(x) > 0 \text{ and } f_o(x) = g_o(x)\}) = 0,$$

the mapping $(f,g) \to f_g$ of $\mathcal{L}^p_+ \times \mathcal{L}^p_+$ into $\mathcal{L}^p_+$ is continuous at (f_o,g_o) with respect to the $\mathcal{L}^p$-seminorm.

<u>Proof</u>: Suppose there exist sequences (f_n), (g_n) in $\mathcal{L}^p_+$ and $\varepsilon > 0$ such that

$$\lim_{n\to\infty}\| f_n - f_o\|_p = 0 = \lim_{n\to\infty} \|g_n - g_o\|_p \quad \text{and}$$

$$\| (f_n)_{g_n} - (f_o)_{g_o}\|_p > \varepsilon \quad \text{for all } n \in \mathbb{N} .$$

Passing to subsequences if necessary, we may assume that

$$\lim_{n\to\infty} f_n = f_o \quad \mu\text{-a.e.} \quad \text{and} \quad \lim_{n\to\infty} g_n = g_o \quad \mu\text{-a.e.}$$

For each $n \in \mathbb{N}$, set

$$A_n := \{x \in \Omega : |(f_o - g_o)(x)| < \frac{1}{n} \quad \text{and} \quad f_o(x) > 0\}.$$

Then (A_n) is a decreasing sequence in $\mathcal{A}$ and

$$\bigcap_{n\in\mathbb{N}} A_n = \{x \in \Omega : f_o(x) = g_o(x) \quad \text{and} \quad f_o(x) > 0\}.$$

Therefore, $\lim\limits_{n\to\infty} \mu(A_n) = \mu(\bigcap\limits_{n\in\mathbb{N}} A_n) = 0$ and $\lim\limits_{n\to\infty}\|f_o\cdot 1_{A_n}\|_p = 0$, where 1_{A_n} denotes the indicator function of A_n in Ω.
Choose $m \in \mathbb{N}$ such that $\|f_o\cdot 1_{A_m}\|_p < \frac{\varepsilon}{3}$ and put

$$A'_m := \{x \in \Omega : f_o(x) > g_o(x) + \frac{1}{m}\}$$

$$A''_m := \{x \in \Omega : f_o(x) < g_o(x) - \frac{1}{m}\} .$$

Since the function $C \to \int_C f_o^p\, d\mu = \|f_o\cdot 1_C\|_p^p$ defines a positive absolutely μ-continuous measure on $\mathcal{A}$ there exists $\delta > 0$ such that $\|f_o\cdot 1_C\|_p < \frac{\varepsilon}{3}$ for all $C \in \mathcal{A}$ satisfying $\mu(C) < \delta$. Egoroff's theorem now ensures that there is an $\mathcal{A}$-measurable set $B \subset \Omega$ and a natural number $n_o \in \mathbb{N}$ such that $\mu(\Omega\setminus B) < \delta$,

$$|(f_n - f_o)(x)| < \frac{1}{2m} \quad \text{and} \quad |(g_n - g_o)(x)| < \frac{1}{2m} \quad \text{for all} \quad x \in B,\ n \geq n_o.$$

Consequently, for all $n \geq n_o$, we obtain

$$(f_n-g_n)(x) = (f_o-g_o)(x)-(f_o-f_n)(x)-(g_n-g_o)(x) > 0, \text{ whenever } x \in B\cap A'_m ,$$

$$(g_n-f_n)(x) = (g_o-f_o)(x)-(g_o-g_n)(x)-(f_n-f_o)(x) > 0, \text{ whenever } x \in B\cap A''_m .$$

Choosing $n \geq n_o$ such that $\|f_n - f_o\|_p < \frac{\varepsilon}{3}$ we conclude

$$\|(f_n)_{g_n} - (f_o)_{g_o}\|_p = \|f_n \cdot 1_{\{f_n \geq g_n\}} - f_o \cdot 1_{\{f_o \geq g_o\}}\|_p$$

$$\leq \|f_n \cdot 1_{\{f_n \geq g_n\}} - f_o \cdot 1_{\{f_n \geq g_n\}}\|_p + \|f_o \cdot 1_{\{f_n \geq g_n\}} - f_o \cdot 1_{\{f_o \geq g_o\}}\|_p$$

$$\leq \|f_n - f_o\|_p + \|f_o \cdot 1_{A_m \cup (\Omega \setminus B)}\|_p \leq \frac{\varepsilon}{3} + \|f_o \cdot 1_{A_m}\|_p + \|f_o \cdot 1_{\Omega \setminus B}\|_p$$

$$< \varepsilon$$

contradicting the assumption. ∎

<u>5.2 Notations</u>: Let $(\Omega, \mathcal{A}, \mu)$ be a finite measure space.

a) Two $\mathcal{A}$-measurable functions $f, g : \Omega \to \mathbb{R}$ will be called <u>μ-almost-nowhere proportional</u> (or, in short, <u>μ-a.n. proportional</u>) if the relation

$$\mu(\{x \in \Omega : \alpha f(x) = \beta g(x)\}) \neq 0$$

implies $\alpha = \beta = 0$ whatever $\alpha, \beta \in \mathbb{R}$ has been choosen.

b) Let $(f_i)_{i \in I}$ be a finite family in $\mathcal{L}_+^p$ such that f_i, f_j are μ-a.n. proportional for every choice $i, j \in I$, $i \neq j$. If $\lambda := (\lambda_i)_{i \in I} \in \mathbb{R}_+^I$ and $g := \bigvee_{j \in I} \lambda_j f_j$ let

$$f_{i,\lambda} := (\lambda_i f_i)_g \quad \text{for each } i \in I.$$

Setting $g_\lambda^{(i)} := \bigvee_{j \in I \setminus \{i\}} \lambda_j f_j$ note that $f_{i,\lambda} = (\lambda_i f_i)_{g_\lambda^{(i)}}$ for all $i \in I$.

Since

$$\mu(\{x \in \Omega : \lambda_i f_i(x) = g_\lambda^{(i)}(x)\}) \leq \sum_{j \in I \setminus \{i\}} \mu(\{x \in \Omega : \lambda_i f_i(x) = \lambda_j f_j(x)\}) = 0$$

whenever $\lambda_i \neq 0$, the functions $\lambda_i f_i$ and $g_\lambda^{(i)}$ satisfy the assumptions of Lemma 5.1 for each $\lambda \in \mathbb{R}_+^I$ and each $i \in I$. Hence the mapping $\lambda \to \|f_{i,\lambda}\|_p$ is continuous on $\mathbb{R}_+^I$.

c) Let I again denote a finite set. For a subset $J \subset I$, $\rho \in \mathbb{R}_+$ and each family $\lambda = (\lambda_i)_{i \in I} \in \mathbb{R}_+^I$ let $\lambda_{\rho,J} = (\eta_i)_{i \in I}$ be defined by

$$\eta_i = \begin{cases} \rho\lambda_i & \text{if} \quad i \in J \\ \lambda_i & \text{for} \quad i \in I \setminus J . \end{cases}$$

To simplify the statements of the next two lemmas we introduce the following function $\varphi : \mathbb{R}_+^I \to \mathbb{R}_+^I$ for a fixed family $(\alpha_i)_{i \in I} \in]0,\infty[^I$ and a real number $q \in [1,\infty[$:

$$\varphi(\lambda) = (\varphi_i(\lambda))_{i \in I} \text{ , where } \quad \varphi_i(\lambda) := \frac{(\lambda_i)^{1/p} \cdot \| f_{i,\lambda} \|_p^{1/q}}{(\alpha_i)^{1/p}}$$

and $(f_i)_{i \in I}$ is a finite family in $\mathscr{L}_+^p$ with μ-a.n. proportional functions f_i, f_j for $j \neq i$.

Finally, we recall the definition of a proper mapping from a Hausdorff topological space X into a locally compact space Y: A continuous mapping $\psi : X \to Y$ is proper (see [16], Ch. I, § 10.3, Prop. 7), if $\psi^{-1}(K)$ is compact for each compact subset $K \subset Y$.

The following lemma now subsumes some important properties of the mapping φ introduced above:

5.3 Lemma: The mapping $\varphi : \mathbb{R}_+^I \to \mathbb{R}_+^I$ is continuous, proper and satisfies the following conditions:

(5.3.1) For every choice of $J \subset I$, $\lambda \in \mathbb{R}_+^I$, $\rho > 0$

$$\left.\begin{array}{ll} \varphi_i(\lambda_{\rho,J}) < \varphi_i(\lambda) & \text{for all } i \in J \text{ such that } \varphi_i(\lambda) > 0 \\ \varphi_i(\lambda_\rho, J) \geq \varphi_i(\lambda) & \text{for all } i \in I \setminus J \end{array}\right\} \text{whenever } \rho < 1$$

$$\left.\begin{array}{ll} \varphi_i(\lambda_{\rho,J}) \geq \varphi_i(\lambda) & \text{for all } i \in J \\ \varphi_i(\lambda_{\rho,J}) \leq \varphi_i(\lambda) & \text{for all } i \in I \setminus J \end{array}\right\} \text{whenever } \rho > 1.$$

Proof: The continuity of φ is an immediate consequence of Lemma 5.1. In order to show that φ is proper, suppose to the contrary that there exists a compact subset $K \subset \mathbb{R}_+^I$ such that $\varphi^{-1}(K)$ is unbounded. Then we can extract a sequence $(\lambda^{(n)})_{n\in\mathbb{N}}$ in $\varphi^{-1}(K)$ such that $\|\lambda^{(n)}\| > n$ for each $n \in \mathbb{N}$, where $\|\cdot\|$ denotes an arbitrary norm on $\mathbb{R}^I$. The unit sphere in $\mathbb{R}^I$ being compact, we may assume that the sequence

$$(\sigma^{(n)})_{n\in\mathbb{N}} := \left(\frac{\lambda^{(n)}}{\|\lambda^{(n)}\|}\right)_{n\in\mathbb{N}}$$

is convergent. Let σ denote its limit. Since $\|\sigma\| = 1$, it follows that

$$0 \neq \|\bigvee_i \sigma_i f_i\|_p^p = \sum_{i\in I} \|f_{i,\sigma}\|_p^p$$

from the relations $f_{i,\sigma} \wedge f_{j,\sigma} = 0$ valid for $i,j \in I$, $i \neq j$, and $\sum_i f_{i,\sigma} = \bigvee_i f_{i,\sigma} = \bigvee_i \sigma_i f_i$. Hence there exists $i_o \in I$ such that $\|f_{i_o,\sigma}\|_p \neq 0$. By the definition of $f_{i_o,\sigma}$ we obtain $\sigma_{i_o} \neq 0$, and the equality

$$\lim_{n\to\infty} \frac{\lambda_{i_o}^{(n)}}{\|\lambda^{(n)}\|} = \sigma_{i_o}$$

yields that $(\lambda_{i_o}^{(n)})_{n\in\mathbb{N}}$ must be unbounded.

Finally, since the mapping $\rho \to f_{i_o,\rho}$ from $\mathbb{R}_+^I$ into $\mathcal{L}_+^p$ is positively homogeneous we find out that

$$\varphi_{i_o}(\lambda^{(n)}) = \frac{(\lambda_{i_o}^{(n)})^{1/p} \|f_{i_o,\lambda^{(n)}}\|_p^{1/q}}{\alpha_{i_o}^{1/p}} = \frac{(\lambda_{i_o}^{(n)})^{1/p} \|\lambda^{(n)}\|^{1/q} \|f_{i_o,\sigma^{(n)}}\|_p^{1/q}}{\alpha_{i_o}^{1/p}}$$

for all $n \in \mathbb{N}$. Thus $(\varphi_{i_o}(\lambda^{(n)}))_{n\in\mathbb{N}}$ is an unbounded sequence contradicting the fact that $(\varphi(\lambda^{(n)}))_{n\in\mathbb{N}}$ lies in the compact set K.

In order to prove 5.3.1, let $J \subset I$, $\lambda \in \mathbb{R}_+^I$, $\rho \in]0,1[$ be given. For each $i \in J$, $x \in \Omega$ the inequality

$$\rho\lambda_i f_i(x) \geq \bigvee_{j\in J} (\rho\lambda_j f_j(x)) \vee \bigvee_{k\in I\setminus J} (\lambda_k f_k(x)) \quad \text{implies}$$

$$\lambda_i f_i(x) \geq \bigvee_{j\in J} (\lambda_j f_j(x)) \vee \frac{1}{\rho} \bigvee_{k\in I\setminus J} (\lambda_k f_k(x)) \geq \bigvee_{j\in I} \lambda_j f_j(x).$$

Consequently, we have the inclusion

$$\{x \in \Omega : \rho\lambda_i f_i(x) \geq \bigvee_{j\in J} (\rho\lambda_j f_j(x)) \vee \bigvee_{k\in I\setminus J} (\lambda_k f_k(x))\} \subset$$

$$\subset \{x \in \Omega : \lambda_i f_i(x) \geq \bigvee_{j\in J} \lambda_j f_j(x)\}$$

which yields $f_{i,\lambda_{\rho,J}} \leq f_{i,\lambda}$.

Similarly, we obtain $f_{i,\lambda_{\rho,J}} \geq f_{i,\lambda}$ for each $i \in I \setminus J$.

From the definition of the functions $\varphi_i (i \in I)$ we hence deduce condition 5.3.1 for $\rho < 1$. For $\rho < 1$ the proof is similar. ∎

5.4 Lemma: Let $\varphi : \mathbb{R}_+^I \to \mathbb{R}_+^I$ be the mapping introduced before Lemma 5.3. For each $i_o \in I$ and for every non-negative real number γ there exists a point $x = (\xi_i)_{i\in I} \in \mathbb{R}_+^I$ such that $\xi_{i_o} = \gamma$ and $\varphi_i(x) = \varphi_j(x)$ for all $i,j \in I$, where φ_i again denotes the i-th component of φ.

Proof: If $M := \{(\xi_i)_{i\in I} \in \mathbb{R}_+^I : \xi_{i_o} = \gamma\}$ and $J := I \setminus \{i_o\}$ we set $B := \{x \in M : \xi_i \leq 1 \text{ for all } i \in J\}$. The mapping φ being continuous and B compact, $\alpha := \sup \varphi_{i_o}(B)$ exists in $\mathbb{R}_+$. The following argument shows that $\alpha = \sup \varphi_{i_o}(M)$: If $x = (\xi_i) \in M \setminus B$, then $d := \max_{i\in J} \xi_i > 1$, which yields $\rho := \frac{1}{d} < 1$. From the definition of $x_{\rho,J}$ and from property 5.3.1 we deduce that $x_{\rho,J} \in B$ and $\alpha \geq \varphi_{i_o}(x_{\rho,J}) \geq \varphi_{i_o}(x)$. Hence $\alpha = \sup \varphi_{i_o}(M)$.

If $i \to \delta_{ii_o}$ denotes the characteristic function of the singleton $\{i_o\}$ in I (Kronecker's delta function) and if $z := (\gamma \cdot \delta_{ii_o})_{i\in I}$, then $z_{\rho,J} = z$ for all $\rho < 1$. Using again property 5.3.1 we obtain $\varphi_i(z_{\rho,J}) = \varphi_i(z) = 0$ for all $i \in J$. Therefore the set

$$K := \{x \in M : \varphi_i(x) \leq 2\varphi_{i_o}(x) \text{ for all } i \in I\}$$

is non-empty and compact as a closed subset of the compact set

$$\{x \in M : \varphi_i(x) \leq 2\alpha \text{ for all } i \in I\} = \varphi^{-1}([0,2\alpha]^I) \cap M.$$

The continuous mapping $\psi : K \to \mathbb{R}_+$ defined by

$$\psi(x) = \max_{i \in I} \varphi_i(x) - \min_{i \in I} \varphi_i(x)$$

thus attains its minimum at some point $x_o \in K$. We shall show that $\psi(x_o) = 0$ which will complete the proof.

As a first step let us check the equality

(5.4.1) $\varphi_{i_o}(x) = \max_{i \in I} \varphi_i(x)$ for all $x \in K$ satisfying $\psi(x) = \min \psi(K)$.

Given $x \in K$ such that $i_o \notin J' := \{i \in I : \varphi_i(x) = \max_{k \in I} \varphi_k(x)\}$ note that $\varepsilon := \max_{i \in I} \varphi_i(x) - \max_{i \in I \setminus J'} \varphi_i(x) > 0$. Since φ is continuous we can find a real number $\rho \in]0,1[$ with the following properties:

$$\varphi_j(x_{\rho,J'}) > \varphi_j(x) - \frac{\varepsilon}{2} \quad \text{for all } j \in J',$$

$$\varphi_i(x_{\rho,J'}) < \varphi_i(x) + \frac{\varepsilon}{2} \quad \text{for all } i \in I \setminus J'.$$

Condition 5.3.1 then yields

$$\varphi_j(x_{\rho,J'}) < \varphi_j(x) \leq 2\varphi_{i_o}(x) \leq 2\varphi_{i_o}(x_{\rho,J'}) \quad \text{for all } j \in J' \text{ and}$$

$$\varphi_i(x) \leq \varphi_i(x_{\rho,J'}) < \varphi_i(x) + \frac{\varepsilon}{2} < \max_{k \in I} \varphi_k(x) \leq 2\varphi_{i_o}(x) \leq 2\varphi_{i_o}(x_{\rho,J'})$$
$$\text{for all } i \in I \setminus J'.$$

Hence $x_{\rho,J'} \in K$ and

$$\psi(x_{\rho,J'}) = \max_{j \in J'} \varphi_j(x_{\rho,J'}) - \min_{i \in I \setminus J'} \varphi_i(x_{\rho,J'}) < \max_{j \in J'} \varphi_j(x) - \min_{i \in I \setminus J'} \varphi_i(x) = \psi(x),$$

which shows that $\psi(x) \neq \min \psi(K)$. Thus (5.4.1) holds.

Suppose that $\psi(x_o) > 0$. The set $K_o := \{x \in K : \psi(x) = \min \psi(K)\}$ being compact as a closed subset of K and containing x_o there exists a maximal upper bound x of x_o in K_o with respect to the product ordering in $\mathbb{R}_+^I$ (cf. [36], page 86, 3.1.16).

From the property 5.4.1 we thus deduce that

$$\varphi_{i_o}(x) = \max_{i\in I} \varphi_i(x) \geq \psi(x) > 0.$$

Hence, $i_o \notin J'' := \{i \in I : \varphi_i(x) = \min_{k\in I}\varphi_k(x)\}$, and

$\varepsilon' := \min_{i\in I\setminus J''} \varphi_i(x) - \min_{i\in I} \varphi_i(x) > 0$. The inequality $2\varphi_{i_o}(x) - \varphi_i(x) \geq$

$\geq \varphi_{i_o}(x) > 0$ being true for all $i \in I$, there exists an open neighborhood U of x in M such that for all $y \in U$

$$\varphi_j(y) < \varphi_j(x) + \frac{\varepsilon'}{2} \qquad (j \in J''),$$

$$\varphi_i(y) > \varphi_i(x) - \frac{\varepsilon'}{2} \qquad (i \in I \setminus J'')$$

$$2\varphi_{i_o}(y) - \varphi_i(y) > 0 \qquad (i \in I).$$

Therefore, $U \subset K$ and, moreover,

$$\varphi_j(y) < \varphi_j(x) + \frac{\varepsilon'}{2} = \min_{i\in I} \varphi_i(x) + \frac{\varepsilon'}{2} = \min_{i\in I\setminus J''} \varphi_i(x) - \frac{\varepsilon'}{2}$$

$$\leq \min_{i\in I\setminus J''} \varphi_i(y) \leq \max_{i\in I\setminus J''} \varphi_i(y) \quad \text{for all } y \in U,\ j \in J''.$$

In particular, $\psi(y) = \max_{i\in I\setminus J''} \varphi_i(y) - \min_{j\in J''} \varphi_j(y)$.

For each $y \in U$, $\rho > 1$ such that $y_{\rho,J''} \in U$ we conclude from (5.3.1):

$$\psi(y_{\rho,J''}) = \max_{i\in I\setminus J''} \varphi_i(y_{\rho,J''}) - \min_{j\in J''} \varphi_j(y_{\rho,J''})$$

$$\leq \max_{i\in I\setminus J''} \varphi_i(y) - \min_{j\in J''} \varphi_i(y) = \psi(y).$$

Since $\lim_{\substack{\rho\to 1\\ \rho>1}} x_{\rho,J''} = x$, there exists $\rho > 0$ such that $x_{\rho,J''} \in U$.

Hence $\psi(x_{\rho,J''}) \leq \psi(x) = \min \psi(K)$, which yields $x_{\rho,J''} \in K_o$. The element $x =: (\xi_i)_{i\in I}$ being a maximal upper bound of x_o in K_o we obtain $x_{\rho,J''} = x$, i.e. $\xi_j = 0$ for all $j \in J''$.

Let $y = (\eta_i)_{i\in I}$ be defined by

$$\eta_i = \begin{cases} 0, & \text{if } i \in I \setminus J'' \\ 1, & \text{if } i \in J''. \end{cases}$$

For every choice of $\sigma, \tau \in]0,\infty[$ such that $x + \sigma y \in U$, $x + \tau y \in U$ and $\sigma < \tau$

we have

$$(x+\sigma y)_{\rho,J''} = x+\tau y,$$

where $\rho := \frac{\tau}{\sigma}$.

Consequently,

$$\psi(x+\tau y) = \psi((x+\sigma y)_{\rho,J''}) \leq \psi(x+\sigma y),$$

which shows that the function $\tau \to \psi(x+\tau y)$ is decreasing on $W := \{\tau \in]0,\infty[: x+\tau y \in U\}$. From the continuity of ψ we thus conclude that

$$\psi(x) = \lim_{\substack{\sigma\to 0\\ \sigma\in W}} \psi(x+\sigma y) \geq \psi(x+\tau y) \quad \text{for all } \tau \in W,$$

or, equivalently, $x+\tau y \in K_0$ for all $\tau \in W$, contradicting the maximality of x. ∎

We are now prepared to prove the main theorem of this section. If $p \in]1,\infty[$, then it follows from the theorems of Kakutani (for $p = 1$) and Ando, Bohnenblust, Nakano (for $p \neq 1$) (cf. [66], Ch. 8.5; [43] page 135, Theorem 3) that every Banach lattice with p-additive norm is isometrically isomorphic to a space $L^p(\mu)$ for some measure space $(\Omega,\mathcal{A},\mu)$. Thus the phrase "let E be a Banach lattice with p-additive norm, $p \in [1,\infty[$" may be replaced by the statement "let $(\Omega,\mathcal{A},\mu)$ be a measure space, $1 \leq p < \infty$, and $E = L^p(\mu)$". In some applications, however, it is more convenient to use the p-additivity directly.

<u>5.5 Theorem</u>: Given $p,q \in]1,\infty[$ such that $1/p+1/q \leq 1$ let E and F be Banach lattices with p- and q-additive norms, respectively. For every finite family $(e_i,f_i)_{i\in I}$ in $E \times F_+$ satisfying $e_i^+ \neq 0$, $f_i \neq 0$ and $f_i \wedge f_j = 0$ for all $i,j \in I$, $i \neq j$, there exists a family $(\lambda_i)_{i\in I} \in]0,\infty[^I$ such that

$$\|\bigvee_{i\in I} \lambda_i e_i^+\|_p \cdot \|\sum_{i\in I} \frac{1}{\lambda_i} f_i\|_q \leq r^{\otimes}(\sum_{i\in I} e_i \otimes f_i),$$

where r is the bisublinear functional on $E \times F$ introduced before Corollary 3.20.

Proof: Since E is isometrically isomorphic to a space $L^p(\mu)$ for some measure space $(\Omega, \mathcal{A}, \mu)$ we may assume that $E = L^p(\mu)$.

Step 1: Let us first assume that the measure μ is finite and that the elements $e_i^+ (i \in I)$ are pairwise μ-a.n. proportional, in the sense that $\bar{e}_i^+ (i \in I)$ are pairwise μ-a.n. proportional for some (and hence any) choice of representatives $\bar{e}_i \in \mathcal{L}^p(\mu)$ in the equivalence classes e_i. If $\lambda = (\lambda_i) \in \mathbb{R}_+^I$ we define $e_{i,\lambda}^+ \in L^p(\mu)$ for each $i \in I$ to be the equivalence class of $\bar{e}_{i,\lambda}^+$ (see 5.2, b), where

$$\bar{e}_{i,\lambda}^+(x) = \begin{cases} \lambda_i \bar{e}_i^+(x) & \text{for all } x \in \Omega \text{ such that } \lambda_i \bar{e}_i^+(x) \geq \bigvee_{j \in I} \lambda_j \bar{e}_j^+(x), \\ 0 & \text{else.} \end{cases}$$

Note that $e_{i,\lambda}^+$ is clearly independent of the particular choice of the functions $\bar{e}_i$ in the equivalence classes e_i.

By Lemma 5.4 there exists a family $\lambda = (\lambda_i)_{i \in I}$ of non-negative real numbers such that $\lambda_{i_o} = 1$ for some $i_o \in I$ and

$$(5.5.1) \qquad \frac{\lambda_i^{1/p} \, \|e_{i,\lambda}^+\|^{1/q}}{\|f_i\|^{1/p}} = \frac{\lambda_j^{1/p} \, \|e_{j,\lambda}^+\|^{1/q}}{\|f_j\|^{1/p}} \quad \text{for all } i,j \in I \quad {}^{+)}$$

If $e_{i,\lambda}^+ = 0$ for some $i \in I$, then from the definition of $e_{j,\lambda}^+$ and from (5.5.1) we obtain $e_{j,\lambda}^+ = 0$ for all $j \in I$. The resulting inequality $e_{i_o}^+ = \lambda_{i_o} \cdot e_{i_o}^+ \leq \bigvee_{j \in I} \lambda_j e_j^+ = \sum_{j \in I} e_{j,\lambda}^+ = 0$, however, contradicts the assumption $e_{i_o}^+ \neq 0$. Thus $\|e_{i,\lambda}^+\| > 0$ and, in particular, $\lambda_i > 0$ for all $i \in I$.

Select positive linear forms e_i', f_i' on E or F, respectively, such that

---

+) The meaning being obvious from the context we shall omit the subscripts p and q of $\|\cdot\|_p$ and $\|\cdot\|_q$, respectively.

$$\|e_i'\| = 1 = \|f_i'\|,$$

$$e_i'(e_{i,\lambda}^+) = \|e_{i,\lambda}^+\|,$$

$$f_i'(f_i) = \|f_i\|,$$

for each $i \in I$. Since $p,q \in]1,\infty[$, the linear forms e_i', f_i' are uniquely determined. Therefore, $e_i' \circ P = e_i'$ and $f_i' \circ Q_i = f_i'$ for the band projections P_i, Q_i from E and F on the bands generated by $e_{i,\lambda}^+$ and f_i, respectively. The families $(e_i')_{i\in I}$ and $(f_i')_{i\in I}$ thus contain pairwise orthogonal linear forms. Setting $q' := q/(q-1)$ we define

$$\alpha_i := \|e_{i,\lambda}^+\|^{(p-q')/q'} \text{ for all } i \in I \text{ and}$$

$$\beta := \|\bigvee_{i\in I} \lambda_i e_i^+\|^{(p-q')/q'}.$$

Then $q' \leq p$ for $1/p + 1/q \leq 1$. Moreover, the norm of F' is q'-additive. We claim that the positive operator $T : E \to F'$ given by

$$Te = \frac{1}{\beta} \sum_i \alpha_i e_i'(e) f_i'$$

is contractive. Indeed, if $q' < p$ and $e \in E_+$ we obtain

$$\|\beta Te\|^{q'} = \sum_i (\alpha_i e_i'(e))^{q'} \leq \sum_i \alpha_i^{q'} \|P_i e\|^{q'}.$$

The real numbers $t := p/q' \in]1,\infty[$ and $t' := p/(p-q') \in]1,\infty[$ satisfying the relation $1/t + 1/t' = 1$ we conclude from Hölder's inequality

$$\sum_i \alpha_i^{q'} \|P_i e\|^{q'} \leq (\sum_i (\alpha_i^{q'})^{t'})^{1/t'} (\sum_i (\|P_i e\|^{q'})^t)^{1/t} =$$

$$= (\sum_i \|e_{i,\lambda}^+\|^p)^{1/t'} \cdot (\sum_i \|P_i e\|^p)^{1/t} =$$

$$= \|\sum_i e_{i,\lambda}^+\|^{p/t'} \cdot \|\sum_i P_i e\|^{p/t} = \|\sum_i e_{i,\lambda}^+\|^{p-q'} \cdot \|(\sum_i P_i)(e)\|^{q'},$$

where we use the p-additivity of the norm on E and the orthogonality of the families $(e_{i,\lambda}^+)_{i\in I}$ and $(P_i e)_{i\in I}$.

Moreover, $\sum_i P_i$ is again a band projection and $\sum_i e_{i,\lambda}^+ = \bigvee_i e_{i,\lambda}^+ = \bigvee_i \lambda_i e_i^+$ by the definition of $e_{i,\lambda}^+$. Consequently,

$$\|\beta Te\|^{q'} \le \|\bigvee_i \lambda_i e_i^+\|^{p-q'} \cdot \|(\sum_i P_i)(e)\|^{q'} \le \beta^{q'} \cdot \|e\|^{q'},$$

which yields $\|Te\| \le \|e\|$.

For $p = q'$ the same inequality immediately results from

$$\|Te\|^{q'} = \|\sum_i e_i'(e) f_i'\|^{q'} = \sum_i e_i'(e)^{q'} \le \sum_i \|P_i e\|^{q'} = \sum_i \|P_i e\|^p$$

$$= \|\sum_i P_i e\|^p = \|(\sum_i P_i)(e)\|^{q'} \le \|e\|^{q'}.$$

Since T was positive and $e \in E_+$ was arbitrary we conclude that $\|T\| \le 1$. From the definition of the band projections P_i $(i \in I)$ we deduce that $P_i(\lambda_i e_i) = e_{i,\lambda}^+$ for all $i \in I$. Hence,

$$\beta r^{\otimes}(\sum_i e_i \otimes f_i) \ge \beta \cdot \sum_i Te_i(f_i) = \sum_i \alpha_i \|f_i\| e_i'(e_i) = \sum_i \alpha_i \|f_i\| e_i'(P_i e_i) =$$

$$= \sum_i \frac{\alpha_i}{\lambda_i} \|f_i\| e_i'(P_i(\lambda_i e_i)) = \sum_i \frac{\alpha_i}{\lambda_i} \|f_i\| e_i'(e_{i,\lambda}^+) = \sum_i \frac{\alpha_i}{\lambda_i} \|f_i\| \; \|e_{i,\lambda}^+\|$$

$$= \sum_i \|\frac{f_i}{\lambda_i}\| \cdot \|e_{i,\lambda}^+\|^{p/q'} .$$

Observing that Hölder's inequality

$$\sum_i \xi_i \eta_i \le (\sum \xi_i^q)^{1/q} \cdot (\sum_i \eta_i^{q'})^{1/q'} \qquad ((\xi_i),(\eta_i) \in \mathbb{R}_+^I)$$

becomes an equality provided that $\eta_j \xi_i^{p-1} = \eta_i \xi_j^{p-1}$ for all $i,j \in I$, we can set

$$\xi_i := \frac{\|f_i\|}{\lambda_i}, \quad \eta_i := \|e_{i,\lambda}^+\|^{p/q'} \quad (i \in I)$$

and conclude from (5.5.1) and an application of Hölder's equality

$$r^{\otimes}(\sum_i e_i \otimes f_i) \ge \frac{1}{\beta} \sum_i \xi_i \eta_i = \frac{1}{\beta}(\sum_i \xi_i^q)^{1/q} \cdot (\sum_i \eta_i^{q'})^{1/q'} = \frac{1}{\beta}(\sum_i \|\frac{f_i}{\lambda_i}\|^q)^{1/q} (\sum_i \|e_{i,\lambda}^+\|^p)^{1/q'}$$

$$= \frac{1}{\beta}\|\sum_i \frac{f_i}{\lambda_i}\| \; \|\sum_i e_{i,\lambda}^+\|^{p/q'} = \frac{1}{\beta}\|\sum_i \frac{f_i}{\lambda_i}\| \; \|\bigvee_i \lambda_i e_i^+\|^{p/q'} = \|\bigvee_i \lambda_i e_i^+\| \; \|\sum_i \frac{f_i}{\lambda_i}\| .$$

S t e p 2: For each $i \in I$, let $(e_i^{(n)})_{n \in \mathbb{N}}$ be a sequence in E satisfying the following conditions:

$$e_i^{(n)+} \neq 0 \quad \text{for all } n \in \mathbb{N}$$

$$\lim_{n\to\infty} e_i^{(n)} =: e_i \quad \text{exists in } E \text{ and } e_i^+ \neq 0 .$$

Furthermore, let $(f_i)_{i\in I}$ be an orthogonal family in $F_+ \setminus \{0\}$ (with the same index set I). If, for each $n \in \mathbb{N}$, there exists a finite family $(\lambda_i^{(n)})_{i\in I}$ of strictly positive real numbers such that $\lambda_{i_n}^{(n)} = 1$ for some $i_n \in I$ and

$$r^{\otimes}(\sum_i e_i^{(n)} \otimes f_i) \geq \| \sum_i \frac{f_i}{\lambda_i^{(n)}} \| \; \| \bigvee_i \lambda_i^{(n)} e_i^{(n)+} \|$$

then the same holds for $\sum_i e_i \otimes f_i$ and a suitable family $(\lambda_i) \in \,]0,\infty[^I$.
Indeed, since I is finite, there exists an index $i_o \in I$ such that $i_n = i_o$ for infinitely many $n \in \mathbb{N}$. Passing to a subsequence, if necessary, we may assume that $i_n = i_o$ for all $n \in \mathbb{N}$. From the continuity of $r^{\otimes}$ with respect to the projective topology on $E \otimes F$ it follows that

$$r^{\otimes}(\sum_i e_i \otimes f_i) = \lim_{n\to\infty} r^{\otimes}(\sum_i e_i^{(n)} \otimes f_i).$$

Hence, if $n \in \mathbb{N}$ is choosen sufficiently large, we obtain

$$\frac{1}{\lambda_i^{(n)}} \|f_i\| \, \|e_{i_o}^{(n)+}\| \leq \| \sum_{j\in I} \frac{f_j}{\lambda_j^{(n)}} \| \; \| \bigvee_{j\in I} \lambda_j^{(n)} e_j^{(n)+} \| \leq r^{\otimes}(\sum_{j\in I} e_j^{(n)} \otimes f_j)$$

$$\leq r^{\otimes}(\sum_{j\in I} e_j \otimes f_j) + 1 ,$$

$$\|f_{i_o}\| \, \|\lambda_i^{(n)} e_i^{(n)+}\| \leq \| \sum_{j\in I} \frac{f_j}{\lambda^{(n)}} \| \; \| \bigvee_{j\in I} \lambda_j^{(n)} e_j^{(n)+} \| \leq r^{\otimes}(\sum_{j\in I} e_j \otimes f_j) + 1 ,$$

for each $i \in I$. Since $\lim_{n\to\infty} e_{i_o}^{(n)+} = e_{i_o}^+ \neq 0$ there exist real numbers $M, M' > 0$ such that $M' < \lambda_i^{(n)} < M$ for all $i \in I$ and all sufficiently large n.

We thus may assume that the sequence $(\lambda^{(n)})_{n\in\mathbb{N}} := ((\lambda_i^{(n)})_{i\in I})_{n\in\mathbb{N}}$ is contained in the compact set $[M',M]^I$. Selecting a convergent subsequence $(\lambda^{(n_k)})_{k\in\mathbb{N}}$ of $(\lambda^{(n)})_{n\in\mathbb{N}}$ and setting $\lambda_i := \lim_{k\to\infty} \lambda_i^{(n_k)}$ for each $i \in I$ we obtain $\lambda_{i_o} = 1$ and

$$\| \bigvee_i \lambda_i e_i^+ \| \, \| \sum_i \frac{f_i}{\lambda_i} \| = \lim_{k\to\infty} \| \bigvee_i \lambda_i^{(n_k)} e_i^{(n_k)+} \| \, \| \sum_i \frac{f_i}{\lambda_i^{(n_k)}} \| \le$$

$$\le \lim_{k\to\infty} r^{\otimes}(\sum_i e_i^{(n_k)} \otimes f_i) = r^{\otimes}(\sum_i e_i \otimes f_i) .$$

S t e p 3: We are now able to complete the proof at least for a measure space $(\Omega, \mathfrak{A}, \mu)$ where $\Omega = [a,b]$, $a,b \in \mathbb{R}$, $a < b$, $\mathfrak{A}$ being the σ-algebra of all Lebesgue measurable subsets of $[a,b]$ and μ denoting the Lebesgue measure on $[a,b]$. To show this we assume for simplicity that I is an initial segment $I = \{1,\dots,m\}$ of $\mathbb{N}$. For each equivalence class $e_i \in L^p(\mu)$ $(i \in I)$ choose a representative $g_i \in \mathcal{L}^p(\mu)$ and a sequence $(g_{i,n})_{n\in\mathbb{N}}$ of $\mathfrak{A}$-step functions converging to g_i in $\mathcal{L}^p(\mu)$. If we set

$$g_i^{(n)}(x) = g_{i,n}(x) + \frac{x^i}{n} \qquad (x \in [a,b], n \in \mathbb{N}),$$

then $g_i^{(n)}$, $g_j^{(n)}$ are μ-a.n. proportional for all $i,j \in I$, $i \neq j$, and each $n \in \mathbb{N}$. Indeed, given $n \in \mathbb{N}$, a finite partition $(A_k)_{k\in K}$ of $[a,b]$ into $\mathfrak{A}$-measurable subsets and $(\alpha_k)_{k\in K}$, $(\beta_k)_{k\in K} \in \mathbb{R}^K$ such that

$$g_{i,n} = \sum_{k\in K} \alpha_k 1_{A_k} \qquad \text{and}$$

$$g_{j,n} = \sum_{k\in K} \beta_k 1_{A_k} ,$$

consider the set $S := \{x \in A_k : \alpha g_i^{(n)}(x) = \beta g_j^{(n)}(x)\}$, where $k \in K$, $(\alpha,\beta) \in \mathbb{R}^2 \setminus \{(0,0)\}$ are fixed
Being a subset of the solutions of the non-trivial algebraic equation

$$\alpha x^i - \beta x^j + n(\alpha\alpha_k - \beta\beta_k) = 0$$

the set S is finite and, in particular, it is μ-negligible. Hence $g_i^{(n)}$ and $g_j^{(n)}$ are μ-a.n. proportional.

For each $i \in I$ and $n \in \mathbb{N}$, let $e_i^{(n)} \in L^p(\mu)$ be the equivalence class of $g_i^{(n)} \in \mathcal{L}^p(\mu)$. Then $\lim_{n\to\infty} e_i^{(n)} = e_i$ in $L^p(\mu)$ for each $i \in I$. Thus the assertion follows from the first two steps of the proof.

S t e p 4: Finally, let $(\Omega, \mathcal{A}, \mu)$ be an arbitrary measure space and let $g_i \in \mathcal{L}^p(\mu)$ be a function in the equivalence class $e_i \in L^p(\mu)$ for each $i \in I$. If $\mathcal{B}$ is the smallest sub-σ-algebra of $\mathcal{A}$ such that all functions $g_i\,(i \in I)$ are $\mathcal{B}$-measurable and if ν denotes the restriction of μ to $\mathcal{B}$ then $e_i \in L^p(\nu)$ for all $i \in I$. Moreover, $E_o := L^p(\nu)$ is separable and the conditional expectation operator $P : L^p(\mu) \to L^p(\nu)$ is a positive, contractive projection. Consequently, if we set

$$r_o^{\otimes}(t) := \sup\{T_o^{\otimes}(t) : T_o : E_o \to F' \text{ positive, contractive}\},$$

we obtain

$$r^{\otimes}(\sum_i e \otimes f_i) = \sup\{\sum_i Te_i(f_i) : T : E \to F' \text{ positive, contractive}\}$$

$$\geq \sup\{\sum_i (T_o \circ P)(e_i)(f_i) : T_o : E_o \to F' \text{ positive,contractive}\}$$

$$= r_o^{\otimes}(\sum_i e_i \otimes f).$$

We may therefore assume without loss of generality that $E = L^p(\mu)$ is separable. In particular, $(\Omega, \mathcal{A}, \mu)$ is σ-finite. Hence there exists an increasing sequence $(\Omega_n)_{n\in\mathbb{N}}$ of $\mathcal{A}$-measurable sets such that $\mu(\Omega_n) < \infty$ for all $n \in \mathbb{N}$ and

$$\bigcup_{n\in\mathbb{N}} \Omega_n = \Omega .$$

Let $P_n : L^p(\mu) \to L^p(\mu)$ denote the band projection onto the band generated by 1_{Ω_n} in $L^p(\mu)$. Since $\lim_{n\to\infty} P_n e_i = e_i$ for each $i \in I$ and since the band $P_n(L^p(\mu))$ is isometrically isomorphic to $L^p(\mu_n)$, where μ_n denotes the finite measure induced by μ on Ω_n, we may assume that μ is finite using the argument of step 2. Moreover, there exists a measure space $(\Omega', \mathcal{A}', \mu')$ without atoms such that $L^p(\mu)$ can be isometrically imbedded into $L^p(\mu')$ where $L^p(\mu')$ remains separable and μ' has the same total mass as μ (see [52], page 98-99).

In this case, however, $L^p(\mu')$ is isometrically isomorphic to $L^p(\lambda)$, λ denoting the Lebesgue measure on some compact interval $[a,b] \subset \mathbb{R}$

(cf. [43], Ch. 5, § 14, Cor. of Theorem 9). Hence the general case of the assertion is reducible to step 3 of the proof. ∎

5.6 Corollary: Given $p,q \in]1,\infty[$ such that $q \leq p$, let E and G be Banach lattices with p- and q-additive norms, respectively. Then (E,G) is an adapted pair of Banach lattices.

Proof: Property (4.1.1) is an immediate consequence of the reflexivity of G.
The topological dual $F := G'$ of G has q'-additive norm, where $q' = \frac{1}{q-1}$ (cf. [43],Ch. V, § 15, Theorem 2). Note that $1/p + 1/q' \leq 1$ since $q \leq p$. Hence property (4.1.3) follows from Theorem 5.5.
Finally, for a suitable positive measure μ, $L^{q'}(\mu) \cong G' = F$. The space of all equivalence classes of simple functions in $L^{q'}(\mu)$ being a dense vector sublattice of $L^{q'}(\mu)$ condition (4.1.2) is also satisfied. ∎

Corollary 5.6 considerably increases the scope of applications of the theorems 4.4, 4.7, 4.10 and 4.11 of the last section. As a consequence we immediately obtain the following well-known theorem of T. Ando (cf. [5]):

5.7 Theorem: Given $p \in [1,\infty[$ and a measure space $(\Omega, \mathfrak{A}, \mu)$ let H be a closed vector sublattice of $L^p(\mu)$. Then there exists a positive contractive projection $P : L^p(\mu) \to H$.

Proof: H being a Banach lattice with p-additive norm the assertion at once follows from Theorem 4.4 setting $G = H$ and $T : H \to H$ the identity operator. ∎

As a further application we characterize all positively complemented subspaces of L^p-spaces:

5.8 Theorem: (Characterization of the positively complemented subspcaes of L^p-spaces):
Given $p \in [1,\infty[$ and a measure space $(\Omega, \mathfrak{A}, \mu)$ let H be a closed vector subspace of $E := L^p(\mu)$. Then the following are equivalent:

i) There exists a positive projection P from E onto H.

ii) H is a lattice in the ordering induced by E on H and every subset $A \subset H$ bounded from above in E is bounded from above in H.

iii) H is a lattice in the ordering induced by E on H and there is a real number $M > 0$ such that $\| \mathop{\overset{H}{\vee}}\limits_i h_i^{\#} \| \le M \| \bigvee\limits_i h_i^+ \|$ for every finite family $(h_i)_{i \in I}$ in H. Here $h_i^{\#}$ denotes the positive part of h_i in H, while $\mathop{\overset{H}{\vee}}\limits_i h_i$ is the supremum of the finite family $(h_i)_{i \in I}$ in H.

Proof: (i) $\Rightarrow$ (ii): If $P : E \to E$ is a positive projection and $A \subset H := P(E)$ is bounded from above by a in E, then H is a lattice (cf. [66], page 214) and A is bounded from above by Pa in H.

(ii) $\Rightarrow$ (iii): Suppose that (ii) holds, but, for each $n \in \mathbb{N}$, there exists a finite family $(h_i^{(n)})_{i \in I_n}$ in H such that

$$\|a_n\| > n \cdot 2^n \|b_n\|,$$

where $a_n := \mathop{\overset{H}{\vee}}\limits_{i \in I_n} h_i^{(n)\#}$, $b_n := \bigvee\limits_{i \in I_n} h_i^{(n)+}$. Setting $k_n := \dfrac{b_n}{2^n \|b_n\|}$ $(n \in \mathbb{N})$ the series $\sum\limits_{n=1}^{\infty} k_n$ converges absolutely in E and $b := \sum\limits_{n=1}^{\infty} k_n$ is an upper bound in E for every k_n $(n \in \mathbb{N})$. Thus b is also an upper bound in E for the set

$$A := \{\frac{h_i^{(n)}}{2^n \|b_n\|} : n \in \mathbb{N},\ i \in I_n\}.$$

Let a be an upper bound for $A \cup \{0\}$ in H. Then

$$\frac{\mathop{\overset{H}{\vee}}\limits_{i \in I_n} h_i^{(n)\#}}{2^n \|b_n\|} \le a \qquad \text{for all } n \in \mathbb{N},$$

hence

$$n < \frac{\|a_n\|}{2^n\|b_n\|} \le \|a\| \qquad \text{for all } n \in \mathbb{N},$$

which is absurd.

(iii) ⇒ (i): If $(h_i)_{i\in I}$ is a finite family in H_+ such that $h_i \overset{H}{\wedge} h_j = 0$ for all $i,j \in I$, $i \neq j$ ($\overset{H}{\wedge}$ denoting the infimum in H), then

$$\|\sum_i h_i\| = \|\overset{H}{\bigvee_i} h_i\| \le M\|\bigvee_i h_i\| \le M(\sum_i \|h_i\|^p)^{1/p}.$$

Thus, if we define

$$\|h\|_H := \inf\{(\sum_i \|h_i\|^p)^{1/p} : (h_i)_{i\in I} \text{ finite family in } H_+,\ \sum_i h_i = |h|_H,$$
$$h_i \overset{H}{\wedge} h_j = 0 \text{ for all } i,j \in I,\ i \neq j\},$$

where $|h|_H$ is the absolute value of h <u>in H</u>, we obtain

$$\frac{1}{M}\|h\| \le \|h\|_H \le \|h\|.$$

Note that H is Dedekind complete, since an increasing net $(h_i)_{i\in I}$ in H_+ bounded from above in H has a limit in E which necessarily belongs to the closed subspace H. Thus,

$$\|h\|_H = \inf\{(\sum_i \|P_i(|h|_H)\|^p)^{1/p} : (P_i)_{i\in I} \text{ orthogonal family of band}$$
$$\text{projections \underline{in H}},\ \sum_i P_i = \text{identity on } H\}.$$

From this equality we at once deduce that $\|h\|_H$ is a p-additive norm on H making H a Banach lattice. Moreover, for every finite family (h_i) in H we obtain

$$\|\overset{H}{\bigvee_i} h_i^+\|_H \le \|\overset{H}{\bigvee_i} h_i^+\| \le M\|\bigvee_i h_i^+\|.$$

Keeping in mind Theorems 4.4, 4.3 and 4.7 we see that the identity mapping on H has a positive extension $P : E \to H$ such that

$$\frac{1}{M}\|Pe\| \le \|Pe\|_H \le M\cdot\|e\| \qquad \text{for all } e \in E. \quad ■$$

For $p \in]1,\infty[$, L^p-spaces are reflexive. Hence we obtain the following stronger modification of Corollary 4.12 for L^p-spaces, $p \in]1,\infty[$, as an immediate consequence of Theorems 4.10 and 5.6:

5.9 Theorem: Given $p,q \in]1,\infty[$, $q \leq p$, let E and G be Banach lattices with p- and q-additive norms, respectively. Furthermore, let $F := G'$ be the topological dual of G and $w : E \times F \to \mathbb{R}_\infty$ be a bisublinear, l.s.c. functional satisfying the following conditions:

i) $\{f \in F : w(e,f) < \infty\} \subset F_+$ for all $e \in E \setminus \{0\}$,

ii) $f \to w(e,f)$ is additive on F_+ for each $e \in E$,

iii) $e \to w(e,f)$ is isotone for each $f \in F_+$.

Then, for each pair $(e,f) \in E \times F_+$ and every real number α such that

$$-w(-e,f) < \alpha < w(e,f),$$

there exists a w-dominated operator $T : E \to G$ satisfying $f(Te) = \alpha$. ■

Extension theorems for operators from AM-spaces into Dedekind complete vector lattices are usually proved by means of the vector-valued versions of the Hahn-Banach theorem. Since, in this context, many results on operator extension can be found in the mathematical literature, we did not explicitly deal with this case until now. For completeness, the following final example of adapted Banach spaces concerning AM-spaces is added to this section:

5.10 Theorem: Let E be an AM-space and F an arbitrary Banach lattice. Then condition (4.1.3) is satisfied. If G is a Banach lattice such that there is a positive, contractive projection from G" onto the natural image J(G) of G in G" and that G' has a topological orthogonal system[+)], then (E,G) is an adapted pair.

+) see [66], page 169, for the terminology!

Proof: Given a finite family $(e_i,f_i)_{i\in I}$ in $E \times F_+$ such that $e_i^+ \neq 0$, $f_i \neq 0$ and $f_i \wedge f_j = 0$ for all $i,j \in I$, $i \neq j$, we set $\lambda_i := \frac{1}{\|e_i^+\|}$ for each $i \in I$.

Then there exists a positive linear form f' on F satisfying $\|f'\| = 1$ and $f'(\sum_i \frac{1}{\lambda_i} f_i) = \|\sum_i \frac{1}{\lambda_i} f_i\|$.

If, for each $i \in I$, P_i denotes the band projection from F' onto the band $\{\ell \in F' : |\ell| \wedge |\ell'| = 0$ for all $\ell' \in F'$ such that $|\ell'|(f) = 0\}$ (the so-called band of strict positivity of f_i, see [66], page 79), then $f'(f_i) = P_i(f')(f_i)$. Finally, we can select a positive linear form e_i' on E satisfying $\|e_i'\| = 1$ and $e_i'(e_i) = \|e_i^+\| = \frac{1}{\lambda_i}$, for each $i \in I$.

The positive operator $T : E \to F'$, defined by

$$Te = \sum_i e_i'(e) \cdot P_i(f')$$

is contractive, since the relation

$$\|Te\| = \|\sum_i e_i'(e) \cdot P_i(f')\| \leq \|e\| \; \|\sum_i P_i(f')\| \leq \|e\| \; \|f'\| = \|e\|$$

holds for each $e \in E_+$. Consequently,

$$\|\bigvee_i \lambda_i e_i^+\| \; \|\sum_i \frac{1}{\lambda_i} f_i\| = \|\sum_i \frac{1}{\lambda_i} f_i\| = f'(\sum_i \frac{1}{\lambda_i} f_i) = \sum_i \frac{1}{\lambda_i} P_i(f')(f_i)$$

$$= \sum_i Te_i(f_i) \leq r^{\otimes}(\sum_i e_i \otimes f_i) ,$$

which yields (4.1.3).

Furthermore, if the topological dual G' of the Banach lattice G possesses a topological orthonormal system A, then for each $a \in A$ the vector lattice ideal F_a generated by a in $F := G'$ is a Dedekind complete AM-space with unit a. Hence in this case (4.1.2) follows using the same arguments as in the proof of 4.3 and (E,G) is an adapted pair of Banach lattices if condition (4.1.1) is also satisfied. ∎

Concerning norm-preserving linear extensions of non-positive operators in L^p-spaces the tensor product approach outlined in this section might also be successful. In particular, it should be possible to reprove the remarkable results of [50] and [51].

6. The Korovkin closure for equicontinuous nets of positive operators

While until now only a few examples were added to illustrate the efficiency of some results, the last three sections are concerned with applications of the extension theory to convergence problems for nets of positive operators. Since the reader might wonder how to apply the rather inaccessible uniqueness characterizations of the previous sections the results of section 6 - 8 should be instructive even for those who are not interested in theorems of Korovkin type.

Throughout this section E and G will denote Banach lattices specified occasionally, H will be a linear subspace of E and $S : E \to G$ will indicate a vector lattice homomorphism.

6.1 Notations:

a) The class of all nets of positive operators from E into G will be denoted by $\mathcal{P}$, likewise $\mathcal{P}_e$ stands for the class of all equicontinuous nets in $\mathcal{P}$ and $\mathcal{P}_o$ (resp. $\mathcal{P}_o''$) is the class of all nets $(T_i)_{i\in I}$ such that $T_i = P$ for all $i \in I$ and some fixed positive operator $P : E \to G$ (resp. $P : E \to G''$).

b) For an arbitrary class $\mathcal{D}$ of nets of positive operators from E into G (resp. into G") we denote by $\mathrm{Kor}_{\mathcal{D},S}(H)$ the set of all $e \in E$ satisfying the following condition:

If $(T_i)_{i\in I} \in \mathcal{D}$ is such that $\lim_{i\in I} T_i h = Sh$ for all $h \in H$ then $\lim_{i\in I} T_i e = Se$.

$\mathrm{Kor}_{\mathcal{D},S}(H)$ is called Korovkin closure (or shadow) of H with respect to $\mathcal{D}$ and S. If $J : G \to G''$ denotes the natural imbedding, we write

$\mathrm{Kor}_{P_o'',S}(H)$ instead of $\mathrm{Kor}_{P_o'',J\circ S}(H)$.

If $E = G$ and S is the identity mapping we omit the subscript S writing simply $\mathrm{Kor}_{P_o}(H)$, $\mathrm{Kor}_{P_e}(H)$ etc.

c) For $e \in E$ and $\varepsilon > 0$ let

$$\hat{H}_{e,\varepsilon} := \{ \bigwedge_i h_i : (h_i) \in H_{e,\varepsilon} \},$$

where $H_{e,\varepsilon}$ is the set defined in Example 3.21. Then $T := (J \circ S)|_H$ is a positive operator from H into $F' := G''$. Using the notations of 3.21 and observing that S is a vector lattice homomorphism it follows that

$$\hat{T}(e,f) = \sup_{\varepsilon > 0} \inf_{k \in \hat{H}_{e,\varepsilon}} f(Sk) \qquad \text{for all } e \in E,\ f \in F_+ .$$

By Lemma 3.23 and 3.13 the mapping $f \to \hat{T}(e,f)$ may be considered as an element of the sup-completion F'_s of F'. The sublinear mapping $\Phi : E \to F'$ defined by

$$\Phi(e)(f) = \hat{T}(e,f) \qquad (e \in E,\ f \in F_+)$$

will be denoted by $\hat{S}_H$ in the following.

d) By Lemma 3.13, the lattice cone of all l.s.c., additive $\mathbb{R}_\infty$-valued, positively homogeneous functionals on F_+ is isomorphic to the sup-completion of F'. In the same way, the sup-completion C of F' with respect to the inverse (= dual) ordering, i.e. the inf-completion of F', can be identified with the cone of all upper semi-continuous additive, positively homogeneous functionals on F_+ with values in $\mathbb{R} \cup \{-\infty\}$. For each $e \in E$ we define $\check{S}_H(e) \in C$ by the equation

$$\check{S}_H(e)(f) = \inf_{\varepsilon > 0} \sup_{k \in \check{H}^{e,\varepsilon}} f(Sk) \qquad (f \in F_+)$$

where $\check{H}^{e,\varepsilon}$ is the set of all suprema of finite, non-empty subsets A of H such that $\|(e - \sup A)^-\| \leq \varepsilon$.

e) We recall that $\hat{S}_H$ is <u>regularized</u>, if $\hat{S}_H$ coincides with its regularization $\hat{S}_H^{\cap}$. By Lemma 3.26 this is true iff

$$\hat{S}_H(e)(f) = \sup_{\varepsilon>0} \inf_{(h_i,f_i)\in H^{\otimes}_{e,f,\varepsilon}} \sum_i f_i(Sh_i) \quad \text{for all } e\in E,\ f\in F_+ .$$

6.2 Remark: Since P_o (resp. P_o'') contains only constant nets, $\mathrm{Kor}_{P_o,S}(H)$ (resp. $\mathrm{Kor}_{P_o'',S}(H)$) is the set of all $e\in E$ such that

$Te = Se$ for all positive operators $T : E \to G$ (resp. $T : E \to G''$) satisfying $T|_H = S|_H$.

In Theorem 6.7 below we shall provide a first characterization of the respective Korovkin closures. Although this description is, in general, far from being easily applicable it is the starting point for a list of equivalent conditions supplied later on and becoming more practical. The general assumption upon S will be that $\hat{S}_H$ must be regularized By Theorem 4.10 this is true at least when the pair (E,G) of Banach lattices satisfies condition (4.1.3) for $F := G'$. As indicated by the results of the preceding sections most situations in which Korovkin theorems naturally occur are covered by Theorem 4.10. Moreover, the following two lemmas show that $\hat{S}_H$ is regularized also in many cases where condition (4.1.3) fails.

6.3 Lemma: Let G_o be a vector sublattice of G such that $S(H) \subset G_o$ and that $\widehat{J_o \circ S|_H}$ is regularized, where $J_o : G_o \to G_o''$ is the natural imbedding. Then $\hat{S}_H$ is regularized.

Proof: If $F := G'$, then the following relation holds for each $e\in E$, $f\in F_+$:

$\sup\{Te(f) : T : E \to G''$ positive linear extension of $J\circ S|_H\}$
$\geq \sup\{T_o e(f|_{G_o}) : T_o : E \to G_o''$ positive linear extension of $J_o \circ S|_H\}$

$$= \widehat{J_o \circ S|_H}(e)(f|_{G_o}) = \sup_{\varepsilon>0} \inf_{k\in\hat{H}_{e,\varepsilon}} f|_{G_o}(Sk) = \sup_{\varepsilon>0} \inf_{k\in\hat{H}_{e,\varepsilon}} f(Sk) = \hat{S}_H e(f).$$

Hence $\hat{S}_H$ is regularized. ∎

6.4 Corollary: If $S(H)$ is a finite dimensional vector sublattice of G, then $\hat{S}_H$ is regularized.

Proof: Since $S(H)$ is finite dimensional it can be endowed with an L^1-norm equivalent to the norm induced by G on $S(H)$.
Then $(E,S(H))$ is an adapted pair of Banach lattices by Theorem 4.3.
The assertion is thus a consequence of Lemma 6.3. ∎

6.5 Lemma: Let $(P_i)_{i\in I}$ be an increasing net of band projections in $F := G'$ satisfying

$$\lim_{i\in I} P_i f = f \qquad \text{for all } f\in F.$$

If $\widehat{P_i'\circ J\circ S|_H}$ is regularized for each $i\in I$, then $\hat{S}_H$ is regularized.

Proof: For each pair $(e,f)\in E\times F_+$ we obtain

$$(\widehat{P_i'\circ J\circ S|_H})(e)(f) = \sup_{\varepsilon>0} \inf_{k\in\hat{H}_{e,\varepsilon}} P_i'((J\circ S|_H)(k))(f) =$$

$$= \sup_{\varepsilon>0} \inf_{k\in\hat{H}_{e,\varepsilon}} (J\circ S|_H)(k)(P_i f) = \hat{S}_H e(P_i f).$$

Hence it follows from Lemma 3.15 that

$$\lim_{i\in I}(\widehat{P_i'\circ J\circ S|_H})(e)(f) = \lim_{i\in I} \hat{S}_H e(P_i f) = \hat{S}_H e(f).$$

Given $\alpha < \hat{S}_H e(f)$, there exists $i\in I$ such that

$$(f - P_i f)(Se) + (\widehat{P_i' \circ J \circ S|_H})(e)(f) > \alpha .$$

Consequently,

$$(\widehat{P_i' \circ J \circ S|_H})(e)(P_i f) = \hat{S}_H e(P_i(P_i f)) = \hat{S}_H e(P_i f) = (\widehat{P_i' \circ J \circ S|_H})(e)(f) >$$
$$> \alpha - (f - P_i f)(Se).$$

The mapping $\widehat{P_i' \circ J \circ S|_H}$ being regularized we can find a $\widehat{P_i' \circ J \circ S|_H}$-dominated operator $T : E \to F'$ such that

$Te(P_i f) > \alpha - (f-P_i f)(Se)$, or equivalently $(P_i' \circ T + (Id - P_i)' \circ J \circ S)(e)(f) > \alpha$

where Id denotes the identity mapping on F. Moreover, for each $(e',f') \in E \times F_+$ we have the estimate

$$\begin{aligned}(P_i' \circ T + (Id - P_i)' \circ J \circ S)(e')(f') &= Te'(P_i f') + (f' - P_i f')(Se') \\ &\leq \hat{S}_H e'(P_i(P_i f)) + \hat{S}_H e'(f' - P_i f') \\ &= \hat{S}_H e'(P_i f) + \hat{S}_H e'(f' - P_i f') = \hat{S}_H e'(f')\end{aligned}$$

Therefore, $P_i' \circ T + (Id - P_i)' \circ J \circ S$ is $\hat{S}_H$-dominated. Since $\alpha < \hat{S}_H e(f)$ was arbitrary, $\hat{S}_H$ is regularized. ∎

For $q \leq p$ the following result is an obvious consequence of Theorem 4.10:

<u>6.6 Corollary</u>: Given $p,q \in [1,\infty[$ and a measure space $(\Omega, \mathcal{A}, \mu)$ let $S : \ell^p \to L^q(\mu)$ be a vector lattice homomorphism. Then $\hat{S}_H$ is regularized for every linear subspace H of ℓ^p.

<u>Proof</u>: If $q = 1$, the assertion immediately follows from Theorems 4.3 and 4.10. Hence we may assume that $q \neq 1$.
The band generated by $S(\ell^p)$ in $L^q(\mu)$ is again an L^q-space. By Lemma 6.3 we may therefore assume that this band coincides with $L^q(\mu)$. Set-

ting $F := L^q(\mu)'$ we denote by $P_n : F \to F$ the band projection onto the band

$$\{f \in F : |f|(Se_i) = 0 \text{ for all } i \in \mathbb{N} \text{ such that } i \geq n\},$$

where $e_i := (\delta_{ij})_{j\in\mathbb{N}} \in \ell^p$. Then $(P_n)_{n\in\mathbb{N}}$ is an increasing sequence of positive operators. Furthermore, for each $x \in \ell^p$, $x = (\xi_i)_{i\in\mathbb{N}} \geq 0$, and each $f \in F_+$ we have the equality

$$\sup_{n\in\mathbb{N}} P_n f(Sx) = \sup_{n\in\mathbb{N}} f(\sum_{i=1}^{n} \xi_i Se_i) = f(\sum_{i=1}^{\infty} \xi_i Se_i) = f(Sx) .$$

$L^q(\mu)$ being the band generated by $S(\ell^p)$ it follows that

$$\sup_{n\in\mathbb{N}} P_n f(g) = f(g) \qquad \text{for all} \quad g \in L^q(\mu) ,$$

i.e. $\sup_{n\in\mathbb{N}} P_n f = f$. The order continuity of the norm on $L^q(\mu)'$ thus yields $\lim_{n\to\infty} P_n f = f$.

To complete the proof it suffices to show that $\widehat{P_n' \circ J \circ S|_H}$ is regularized for each $n \in \mathbb{N}$, by Lemma 6.5. Let Q_n denote the band projection of $L^q(\mu)$ onto the band generated by $\{Se_1,\ldots,Se_n\}$. Then for $e \in E$, $f \in F_+$ the following equality holds:

$$\widehat{P_n' \circ J \circ S|_H}(e)(f) = \sup_{\varepsilon>0} \inf_{k\in\hat{H}_{e,\varepsilon}} P_n f(Sk) = \sup_{\varepsilon>0} \inf_{k\in\hat{H}_{e,\varepsilon}} P_n f(Q_n(Sk))$$

$$= \sup_{\varepsilon>0} \inf_{k\in\hat{H}_{e,\varepsilon}} f((Q_n \circ S)(k)) = \widehat{(Q_n \circ S)}_H(e)(f) .$$

Consequently, $\widehat{P_n' \circ J \circ S|_H} = \widehat{(Q_n \circ S)}_H$.

Since $(Q_n \circ S)(H) \subset (Q_n \circ S)(\ell^p)$ is contained in the finite dimensional vector sublattice generated by $\{Se_1,\ldots,Se_n\}$ in $L^q(\mu)$, the assertion follows from Corollary 6.4. ∎

Focussing our interest on the original characterization problem for Korovkin closures, the following theorem will be a first step towards a practically satisfactory solution. For the formulation we use the notion of "positive, bounded approximation property" (abbreviated by PBAP). By definition, a Banach lattice B has the PBAP iff there exists an equi-continuous net $(T_i)_{i \in I}$ of positive operators of finite rank on B such that

$$\lim_{i \in I} T_i x = x \quad \text{for all} \quad x \in B .$$

All Banach lattices that are of interest for theorems of Korovkin type, in particular all L^p-spaces and AM-spaces, possess the PBAP (see [66], IV, 2.4).

6.7 Theorem: Let $\hat{S}_H$ be regularized and consider the following statements for $e \in E$:

i) $e \in \mathrm{Kor}_{P_o'', S}(H)$,

ii) $\check{S}_H e = J(Se) = \hat{S}_H e$,

iii) there exist sequences (k_n) and (k_n') of infima and suprema, respectively, of finite subsets of H such that

$$\lim_{n\to\infty} Sk_n = Se = \lim_{n\to\infty} Sk_n' \quad \text{and}$$

$$\lim_{n\to\infty} \| (e - k_n)^+ \| = 0 = \lim_{n\to\infty} \| (e - k_n')^- \| ,$$

iv) $e \in \mathrm{Kor}_{P_e, S}(H)$,

v) $e \in \mathrm{Kor}_{P_o, S}(H)$.

Then the following implications hold:

(i) $\Leftrightarrow$ (ii) $\Rightarrow$ (iii) $\Rightarrow$ (iv) $\Rightarrow$ (v).

If the natural image J(G) of G in G" is a band in G", (i) - (v) are equivalent. The statements (i) - (iv) are equivalent whenever G has the positive bounded approximation property.

Proof: (i) $\Leftrightarrow$ (ii) is an immediate consequence of Theorem 2.12.
(ii) $\Rightarrow$ (iii): Suppose that $J(Se) = \hat{S}_H e$ and that (iii) failed to be

true, i.e., $Se \notin \bigcap_{\varepsilon>0} \overline{S(\hat{H}_{e,\varepsilon})}$. Then there exists an $\varepsilon > 0$ such that

$Se \notin \overline{S(\hat{H}_{e,\varepsilon})}$. Note that $\hat{H} := \{\inf A : A \subset H \text{ finite}, A \neq \emptyset\}$ is a convex cone and then $e + E_+ + U_\varepsilon$ is convex, where $U_\varepsilon := \{e' \in E : \|e'\| \leq \varepsilon\}$. Therefore $\hat{H}_{e,\varepsilon} = \hat{H} \cap (e + E_+ + U_\varepsilon)$ is convex, too. It follows that there exists a continuous linear form $f \in F := G'$ such that

$$f(Sk) \leq 1 < f(Se) \quad \text{for all} \quad k \in \hat{H}_{e,\varepsilon}.$$

f can be written as the difference $f = f_1 - f_2$ of two positive linear forms $f_1, f_2 \in F_+$. For each $\varepsilon' \in]0,\varepsilon[$ we thus obtain

$$f_1(Sk) \leq f_2(Sk) + 1 \quad \text{whenever} \quad k \in \hat{H}_{e,\varepsilon'} \subset \hat{H}_{e,\varepsilon}, \text{ i.e.}$$

$$\inf_{k\in\hat{H}_{e,\varepsilon'}} f_1(Sk) \leq \inf_{k\in\hat{H}_{e,\varepsilon'}} f_2(Sk) + 1.$$

From this inequality we deduce

$$f_1(Se) = \hat{S}_H e(f_1) = \hat{S}_H e(f_2) + 1 = f_2(Se) + 1$$

contradicting the relation $(f_1 - f_2)(Se) > 1$.
Similarly, we conclude $J(Se) = \check{S}_H e$ from $Se \in \bigcap_{\varepsilon>0} \overline{S(\check{H}^{e,\varepsilon})}$.

(iii) $\Rightarrow$ iv): Let $(T_i)_{i\in I} \in \mathcal{P}_e$ be such that $\lim_{i\in I} T_i h = Sh$ for all $h \in H$. Given $\varepsilon > 0$ and $M > 0$ satisfying $\|T_i\| \leq M$ for all $i \in I$ select $k \in \hat{H}_{e,\varepsilon/M}$ such that $\|Sk - Se\| \leq \varepsilon$. If A is a non-empty finite subset of H with infimum $\inf A = k$, then

$$\|(T_i e - Se)^+\| \leq \|(Sk - Se)^+\| + \|(T_i - S)(k)^+\| + \|(T_i(e - k))^+\|.$$

Using the continuity of the lattice operations in G we can find an index $i_1 \in I$ such that

$$\|\inf_{h\in A} T_1 h - \inf_{h\in A} Sh\| \leq \varepsilon \quad \text{for all} \quad i \geq i_1, i \in I.$$

S being a vector lattice homomorphism we deduce

$$\|(T_i k - Sk)^+\| \leq \|(\inf_{h\in A} T_i h - \inf_{h\in A} Sh)^+\| \leq \varepsilon \qquad (i \geq i_1).$$

Consequently, for all $i \geq i_1$,

$$\|(T_i e - Se)^+\| \leq \|Sk - Se\| + \varepsilon + \|T_i\| \, \|(e-k)^+\| \leq 3\varepsilon \, .$$

In the same way observing the membership $Se \in \bigcap_{\varepsilon'>0} \overline{S(\check{H}^{e,\varepsilon'})}$ it follows that there is $i_2 \in I$ such that

$$\|(T_i e - Se)^-\| \leq 3\varepsilon \quad \text{for all} \quad i \geq i_2, \, i \in I.$$

Choosing $i_o \in I$, $i_o \geq i_1, i_2$, we obtain the inequality

$$\|T_i e - Se\| \leq \|(T_i e - Se)^+\| + \|(T_i e - Se)^-\| \leq 6\varepsilon \quad \text{for all} \quad i \geq i_o.$$

Thus, $\lim_{i \in I} T_i e = Se$ which yields that $e \in \mathrm{Kor}_{P_e,S}(H)$.

(iv) ⇒ (v): Note that positive operators between Banach lattices are continuous. Hence, if $T : E \to G$ is a positive operator such that $T|_H = S|_H$ we have $(T_i) \in P_e$ for the trivial net $T_i := T$, $i \in I$, where I is some set consisting of only one element. From (iv) we conclude that $Te = \lim_{i \in I} T_i e = Se$ which yields $e \in \mathrm{Kor}_{P_o,S}(H)$.

If the natural image $J(G)$ of G in G'' is a band then the same argument as in the proof of Theorem 4.11 shows that the implication (v) ⇒ (i) holds.

Finally, suppose that G'' possesses the PBAP. If $e \notin \mathrm{Kor}_{P''_o,S}(H)$, then there is a positive operator $T : E \to G''$ with restriction $T|_H = S|_H$ such that $Te \neq J \circ S(e)$. The subsequent lemma 6.9 supplies a net $(T_i)_{i \in I}$ of positive operators from E into G satisfying the following conditions:

$$\|T_i\| \leq \lambda \|T\| \quad \text{for all } i \in I \text{ and some } \lambda > 1,$$

$$\lim_{i \in I} f(T_i e) = Te(f) \quad \text{for all } f \in G',$$

$$\lim_{i \in I} T_i h = Th = Sh \quad \text{for all } h \in H.$$

In particular, $(T_i e)_{i \in I}$ does not converge to Se. Thus, $e \notin \mathrm{Kor}_{P_e,S}(H)$. ∎

The deduction will be completed by the following two auxiliary results. In order to keep the proofs readable we shall give up the distinction between G and its natural image J(G) in G" identifying G with a linear subspace of G".

6.8 Lemma: If $\mathcal{W}$ denotes the linear space of all continuous operators from E into G" mapping H into G, let $\mathcal{Q}$ be the coarsest locally convex topology on $\mathcal{W}$ such that each of the following maps is continuous:

(i) $T \to Th \quad (h \in H)$,

(ii) $T \to Te(f) \quad (f \in G', e \in E)$.

Then any continuous linear functional on $\mathcal{W}$ is of the form $T \to \sum_i Te_i(f_i)$, where $(e_i, f_i)_{i \in I}$ is a finite family in $E \times G'$.

Proof: On the vector space L of all continuous operators from E into G" let $\sigma(L, E \otimes G')$ be coarsest locally convex topology making all functionals of the form (ii) continuous. Endowing the vector space L_H of all continuous operators from H into G with the strong operator topology the linear subspace

$$\mathcal{W}_o := \{ (T,L) \in L \times L_H : L = T|_H \}$$

of the product space $L \times L_H$ is obviously isomorphic to $\mathcal{W}$ as a topological vector space (choosing e.g. the first projection as a suitable isomorphism). Since the dual of L and L_H is $E \otimes G'$ and $H \otimes G'$, respectively (see [65], Ch. IV, § 4, 4.3, Cor. 4), and since, by the Hahn Banach theorem, every continuous linear form on $\mathcal{W}_o$ is the restriction of a continuous linear functional on the product space $L \times L_H$, the dual $\mathcal{W}_o'$ of $\mathcal{W}_o$ is isomorphic to $(E \otimes G') \otimes (H \otimes G')$. Indeed, the only linear form $\ell \in (E \otimes G') \otimes (H \otimes G')$ vanishing on $\mathcal{W}_o$ is the zero functional, hence $\mathcal{W}_o$ is dense in $L \times L_H$. Observing that $T \to (T, T|_H)$ is an inverse isomorphism from $\mathcal{W}$ onto $\mathcal{W}_o$ we obtain as dual of $\mathcal{W}$

$$\mathcal{W}' = (E \otimes G') + (H \otimes G') = E \otimes G'. \quad \blacksquare$$

If W is the space defined in Lemma 6.8 we obtain

6.9 Lemma: If G" has the PBAP then there exists $\lambda > 1$ and, for each $T \in W$, a net $(T_i)_{i \in I}$ of positive operators from E into G with the following properties:

i) $\|T_i\| \leq \lambda \|T\|$ for all $i \in I$,

ii) $\lim_{i \in I} T_i h = Th$ for all $h \in H$,

iii) $\lim_{i \in I} f(T_i e) = Te(f)$ for all $f \in G'$ and each $e \in E$,

iv) each operator $T_i, i \in I$, has finite rank.

Proof: Let $U \subset E' \otimes G$ and $U" \subset E' \otimes G"$ be the respective closed balls belonging to the operator norm in the space of all operators of finite rank from E into G and from E into G", respectively. (We interpret these operators as elements of $E' \otimes G$ and of $E' \otimes G"$, respectively). Furthermore, let $co(E'_+ \otimes G_+)$ (resp. $co(E'_+ \otimes G"_+)$) be the convex hull of all tensors $e' \otimes g$ (resp. $e' \otimes g"$), where $e' \in E'_+$, $g \in G_+$ (resp. $g" \in G"_+$). It is well-known (see [66], IV, § 4, 4.6, Cor. 1 and IV, § 7, 7.4, Cor.) that the $\sigma(E' \otimes G, E" \otimes G')$-closure of $co(E'_+ \otimes G_+)$ is the cone $(E' \otimes G)_+$ of all positive operators in $E' \otimes G$. In the same manner $(E' \otimes G")_+$ is the closure of $co(E'_+ \otimes G"_+)$ with respect to $\sigma(E' \otimes G", G" \otimes G"')$. This topology being finer then $\sigma := \sigma(E' \otimes G", E" \otimes G')$ $(E' \otimes G")_+$ is a subset of the σ-closure $\overline{co(E'_+ \otimes G"_+)}^{\sigma}$ of $co(E'_+ \otimes G"_+)$. Suppose that there exists a tensor $\sum_i e'_i \otimes g"_i \in \overline{co(E'_+ \otimes G"_+)}^{\sigma} \setminus (E' \otimes G")_+$. Then $\sum_i e'_i(e) g"_i \notin G"_+$ for some $e \in E_+$.
Choosing $g' \in G'_+$ such that $\sum_i e'_i(e) g"_i(g') < 0$ the linear form $\varphi := e \otimes g' \in E"_+ \otimes G'_+$ is non-negative on $co(E'_+ \otimes G"_+)$. Consequently,

$$\sum_i e'_i \otimes g"_i \notin \{t \in E' \otimes G" : \varphi(t) \geq 0\} \supset \overline{co(E'_+ \otimes G"_+)}^{\sigma}$$

contradicting the assumption.

Hence $\overline{\mathrm{co}(E'_+ \otimes G''_+)}^{\sigma} = (E' \otimes G'')_+$.

Furthermore, G_+ being $\sigma(G'',G')$-dense in G''_+ by the bipolar theorem the set $E'_+ \otimes G_+$ is σ-dense in $E'_+ \otimes G''_+$ which yields $\overline{\mathrm{co}(E'_+ \otimes G_+)}^{\sigma} = (E' \otimes G'')_+$.
Note that the set $U_+ := U \cap (E' \otimes G)_+$ is σ-dense in $U''_+ := U'' \cap (E' \otimes G'')_+$.
Indeed, applying the bipolar theorem to the dual pair $(E' \otimes G'', E'' \otimes G')$ we obtain

$$\overline{U}^{\sigma}_+ = U^{\circ\circ}_+ = U^{\circ\circ} \cap (E' \otimes G)^{\circ\circ}_+ = U^{\circ\circ} \cap \mathrm{co}(E'_+ \otimes G_+)^{\circ\circ} =$$

$$= U'' \cap \overline{\mathrm{co}(E'_+ \otimes G_+)}^{\sigma} \qquad \text{(cf. [66], Ch. IV, 5.4, Cor. 3)}$$

$$= U''_+ .$$

Therefore, U_+ is also $\sigma(E' \otimes G'', E \otimes G')$-dense in U''_+.
Using the PBAP of G" each operator of

$$B_+ := \{T \in \mathcal{W} : \|T\| \leq 1,\ T \geq 0\}$$

is a $\sigma(L, E \otimes G')$-clusterpoint of $\lambda U''_+$ for some fixed $\lambda > 1$, where L denotes the space of all continuous operators from E into G". Hence B_+ is contained in the $\sigma(L, E \otimes G')$-closure of λU_+. Finally, $\lambda U_+ \subset \mathcal{W}$ and the closure of the convex set λU_+ is the same for every locally convex topology compatible with the duality $(\mathcal{W}, E \otimes G')$. Thus, Lemma 6.8 yields the assertion. ■

6.10 Remark to Theorem 6.7: The implication (iii) ⇒ (iv) has been proved by several authors (see [38],[73],[26]) in slightly modified form (mostly considering only the identity operator for S). If E = G and S is the identity operator then condition (iii) is equivalent to the relation

$$e \in \overline{\hat{H}} \cap -\overline{\hat{H}},$$

where $\hat{H}$ denotes the cone of infima of all finite subsets of H. This description was first given in [73]. The main point of Theorem 6.7, however, is the fact that an exact characterization of Korovkin closures is given.

On the other hand, Theorem 6.7 is too abstract to provide a quick determination of Korovkin closures. Hence, it will be the aim of the next two sections to find descriptions that work. In ℓ^p-spaces useful characterizations of Korovkin closures can be derived directly from Theorem 6.7:

<u>6.11 Application to ℓ^p-spaces</u> (see [38],[29]): Let H be a linear subspace of ℓ^p, $p \in [1,\infty[$. For each sequence $e \in \ell^p$ set

$$\hat{e}(n) := \sup_{\varepsilon>0} \inf_{\substack{h\in H_+ \\ \|(e-h)^+\|\le\varepsilon}} h(n) \ , \quad \check{e}(n) := \inf_{\varepsilon>0} \sup_{\substack{h\in H \\ \|(e-h)^-\|\le\varepsilon}} h(n) \qquad (n \in \mathbb{N})$$

where h(n) denotes the n-th term of the sequence $h \in \ell^p$. Then the following statements are equivalent for each $e \in \ell^p$:

i) $e \in \mathrm{Kor}_{\mathcal{P}_e}(H)$

ii) $e \in \mathrm{Kor}_{\mathcal{P}_o}(H)$

iii) $\check{e}(n) = e(n) = \hat{e}(n)$ for all $n \in \mathbb{N}$

iv) for each $n \in \mathbb{N}$ and for every positive linear extension $f : \ell^p \to \mathbb{R}$ of $\varepsilon_n|_H$ ε_n denoting the point evaluation at the n-th term the equality $f(e) = e(n)$ holds.

<u>Proof</u>: (i) ⇔ (ii) results from Theorem 6.7, (ii) ⇔ (iii) is a consequence of 2.13.3, iii and (iii) ⇔ (iv) follows from 2.13.3, ii applied to the special case $L^\infty(\mu) = \mathbb{R}$ and $S = \varepsilon_n$. ∎

<u>6.12 Corollary</u>: With the assumptions of 6.11 H is a Korovkin space in ℓ^p, i.e. $\mathrm{Kor}_{\mathcal{P}_e}(H) = \ell^p$ iff for each $n \in \mathbb{N}$ ε_n is the only positive linear extension of $\varepsilon_n|_H$ to ℓ^p.

Thus, e.g., the set

$$\{(\tfrac{1}{n})_{n\in\mathbb{N}} \ , \ (\tfrac{1}{n^2})_{n\in\mathbb{N}} \ , \ (\tfrac{1}{n^3})_{n\in\mathbb{N}}\}$$

generates a Korovkin space $H \subset \ell^p$ of dimension 3 for $1 < p < \infty$. Indeed, for each $n_o \in \mathbb{N}$ the sequence $(\frac{(n-n_o)^2}{n^3})_{n\in\mathbb{N}}$ lies in H which yields $f((\frac{(n-n_o)^2}{n^3})_{n\in N}) = 0$ for every positive linear extension f of $\varepsilon_{n_o}|H$. Consequently, $f = \alpha\cdot\varepsilon_{n_o}$ for some $\alpha \geq 0$, n_o being the only zero of positive sequence $(\frac{(n-n_o)^2}{n})_{n\in\mathbb{N}}$. Finally, $\alpha = 1$, since

$$\frac{1}{n_o} = \varepsilon_{n_o}((\tfrac{1}{n})_{n\in N}) = f((\tfrac{1}{n})_{n\in\mathbb{N}}) = \alpha\cdot\varepsilon_{n_o}((\tfrac{1}{n})_{n\in\mathbb{N}}) = \frac{\alpha}{n_o}. \quad \blacksquare$$

Challenged by the results 6.3-6.6 the reader might ask whether $\hat{S}_H$ is regularized for every vector lattice homomorphism $S: E \to G$, where E and G are arbitrary Banach lattices. The following counterexample, however, demonstrates that this is not true.

6.13 Counterexample: Let $E := L^1(\lambda) \otimes L^\infty(\lambda)$, where λ denotes the Lebesgue measure on [0,1]. If $G := L^\infty(\lambda)$, then the projection S from E onto the second direct summand G is a vector lattice homomorphism. Setting

$$H := \{(f_1,f_2) \in E : f_1 = f_2\}$$

we claim that $\mathrm{Kor}_{P_o'',S}(H) = E$ but $\{e \in E : \check{S}_H e = J(Se) = \hat{S}_H e\} = H$ for the natural imbedding $J: G \to G''$. Since G, and hence also G", has the positive bounded approximation property, $\hat{S}_H$ cannot be regularized. To prove the equation $\mathrm{Kor}_{P_o'',S}(H) = E$, let F denote the dual G' of G. Using the same arguments as in Corollary 2.13.3, ii applied to the sup-completion F_s' of F' instead of the lattice cone C_∞ we obtain:

$$\{e \in E : \inf_{\substack{h\in H \\ \|(e-h)^+\|\leq\varepsilon}} (J\circ S)(h) \leq J(Se) \leq \sup_{\substack{h\in H \\ \|(e-h)^-\|\leq\varepsilon}} (J\circ S)(h) \text{ for all } \varepsilon > 0\}$$

$$= \mathrm{Kor}_{P_o'',S}(H).$$

Given $(f_1,f_2) \in E$ and $x \in [0,1]$ we can find an open interval U_x containing x and a real number $\alpha > \|f_2\|_\infty$ such that

$$\int_{U_x}(f_1 - f_2)^+ d\lambda \le \frac{\varepsilon}{2} \quad \text{and} \quad \int(f_1 - \alpha)^+ d\lambda \le \frac{\varepsilon}{2} .$$

If 1_{U_x} and $1_{[0,1]\setminus U_x}$ denote the equivalence classes of the characteristic functions of U_x and $[0,1] \setminus U_x$ in $[0,1]$, respectively, let $h_x := f_2 \cdot 1_{U_x} + \alpha \cdot 1_{[0,1]\setminus U_x}$. Then $(h_x,h_x) \in H$ and

$$\|((f_1,f_2)-(h_x,h_x))^+\| = \|((f_1-h_x)^+,(f_2-h_x)^+)\| =$$
$$= \max(\|(f_1-h_x)^+\|_1,\|(f_2-h_x)^+\|_\infty) = \|(f_1-h_x)^+\|_1 \le \varepsilon.$$

Because of the compactness of the interval $[0,1]$ there is a finite subset $A \subset [0,1]$ such that $\bigcup_{x\in A} U_x \supset [0,1]$. Since $\inf_{x\in A} h_x = f_2$ in $L^\infty(\lambda)$, we obtain

$$\inf_{x\in A}(J\circ S)(h_x,h_x) = \inf_{x\in A} J(h_x) = J(\inf_{x\in A} h_x) = J(f_2) = (J\circ S)(f_1,f_2).$$

Consequently, for $e := (f_1,f_2)$,

$$\inf_{\substack{h\in H \\ \|(e-h)^+\|\le\varepsilon}} (J\circ S)(h) \le (J\circ S)(e).$$

Similarly, $\sup_{\substack{h\in H \\ \|(e-h)^-\|\le\varepsilon}} (J\circ S)(h) \ge (J\circ S)(e).$

Let us now check the inclusion $\{e \in E : \check{S}_H e = J(Se) = \hat{S}_H e\} \subset H$. The converse inclusion being obvious this will complete the counterexample. Given $e \in E$ such that $\check{S}_H e = J(Se) = \hat{S}_H e$ there exist sequences (k_n) and (k'_n) of infima and suprema, respectively, of finite subsets of H such that

i) $\lim_{n\to\infty} Sk_n = Se = \lim_{n\to\infty} Sk'_n$ and

ii) $\lim_{n\to\infty}\|(e-k_n)^+\| = 0 = \lim_{n\to\infty}\|(e-k'_n)^-\|$.

Since H is a closed vector sublattice of E we deduce that $k_n,k'_n \in H$ for each $n \in \mathbb{N}$. Moreover, if $e = (f_1,f_2)$ for suitable $f_1 \in L^1(\lambda)$, $f_2 \in L^\infty(\lambda)$, then $\lim_{n\to\infty} k_n = (f_2, f_2) = \lim_{n\to\infty} k'_n$ by condition (i) and the definition of H. The resulting equality

$$\lim_{n\to\infty} \|(e-k_n)^-\| = 0 = \lim_{n\to\infty} \|(e-k_n)^+\|$$

shows that $\lim_{n\to\infty} k_n = e \in H$. ∎

In order to simplify the notation we shall no longer distinguish between G and its canonical image in G". The reader is cautioned, however, that this identification requires a strict distinction between suprema of infinite subsets of G formed in G and those formed in G". (The same holds for infima of infinite subsets of G).

6.14 Definitions:

a) For each $a \in E_+$ let

$$E_a := \{e \in E : \exists\, \lambda > 0 : |e| \leq \lambda a\}$$

denote the vector lattice ideal generated by a in E.
Though the equivalence

$$a \lhd b \;\Leftrightarrow\; E_a \subset E_b$$

we define a preordering $\lhd$ on E_+. Note that every countable subset $A \subset E_+$ is bounded from above with respect to this preordering. Indeed, if (a_n) is a sequence such that $A = \{a_n : n \in \mathbb{N}\}$, then

$$a := \sum_{n=1}^{\infty} \frac{a_n}{2^n \max(\|a_n\|, 1)}$$

is a $\lhd$-bound for A.

b) Given a linear subspace H_1 of H and elements $e \in E$, $a \in E_+$,

$$\hat{S}^a_{H_1}(e) := \sup_{\varepsilon>0} \inf_{\substack{h\in H_1 \\ h\geq e-\varepsilon a}} Sh$$

exists in the sup-completion F'_s of F', where F is the dual G' of G.

Here we use the convention $\inf \emptyset = \infty$.

In particular, $\hat{S}^a_H$ and $\hat{S}^a_{H\cap E_a}$ are well-defined (sublinear) mappings.

Similarly, we form

$$\check{S}^a_{H_1}(e) = \inf_{\varepsilon>0} \sup_{\substack{h\in H_1 \\ h\leq e+\varepsilon a}} Sh$$

in the inf-completion C of F' (i.e. the sup-completion of F' with respect to the dual ordering), where $\sup \emptyset = -\infty$, the smallest element of the inf-completion.

Identifying the inf-completion C with all u.s.c., the sup-completion F'_s with all l.s.c., positively homogeneous, additive functionals on F_+ using Lemma 3.13, the inequality

$$\check{S}^a_{H_1}(e) \leq Se \leq \hat{S}^a_{H_1}(e)$$

holds for all $e \in E$.

Given $a,b \in E_+$ such that $a \lhd b$, note that $\hat{S}^a_{H_1}(e) \geq \hat{S}^b_{H_1}(e)$ for each $e \in E$. Similarly, for any choice H_1, H_2 of linear subspaces of H satisfying $H_1 \subset H_2$, we have $\hat{S}^a_{H_1} \geq \hat{S}^a_{H_2}$. In particular, $\hat{S}^a_{H\cap E_a} \geq \hat{S}^a_H$.

The converse inequalities are valid for $\check{S}^a_{H_1}$, $\check{S}^b_{H_1}$, $\check{S}^a_{H_2}$.

c) If H is such that $Kor_{\mathcal{P}_e,S}(H) = E$, H is called a Korovkin space in E. Every subset of H linearly generating H is a Korovkin system (with respect to $\mathcal{P}_e$ and S).

6.15 Theorem: If $\hat{S}_H$ is regularized and if G" has the PBAP or G is a band in G", then the following statements are equivalent for each $e \in E$:

i) $e \in Kor_{\mathcal{P}_e,S}(H)$.

ii) $\hat{S}^a_{H_o\cap E_a}(e) = Se = \check{S}^a_{H_o\cap E_a}(e)$ for some linear subspace $H_o \subset H$ with countable algebraic basis and for some $a \in E_+$

iii) $\check{S}^a_{H\cap E_a}(e) = Se = \hat{S}^a_{H\cap E_a}(e)$ for some $a \in E_+$,

iv) $\check{S}^a_H(e) = Se = \hat{S}^a_H(e)$ for some $a \in E_+$.

Proof: (i) ⇒ (ii): Given $e \in \mathrm{Kor}_{\mathcal{P}_e,S}(H)$, there exists a sequence $(A_n)_{n\in\mathbb{N}}$ of finite subsets of H such that

$$\lim_{n\to\infty} S(\inf A_n) = Se \quad \text{and} \quad \lim_{n\to\infty}\|(e - \inf A_n)^+\| = 0,$$

by Theorem 6,7, iii. Topological convergence being equivalent to relatively uniform $*$-convergence (see [60], Ch. IV, § 2), there is a subsequence $(A_{n_i})_{i\in\mathbb{N}}$ and elements $b \in E_+$, $g \in G_+$ such that

$$\lim_{i\to\infty}\|(e - \inf A_{n_i})^+\|_b = 0 \quad \text{and} \quad \lim_{i\to\infty}\|S(\inf A_{n_i}) - Se\|_g = 0$$

where $\|\cdot\|_b$ and $\|\cdot\|_g$ denote the respective Minkowski functionals of the order intervals $[-b,b]$ and $[-g,g]$. The set $A := \bigcup_{i\in\mathbb{N}} A_{n_i}$ being countable, we can find $c \in E_+$ such that $b \lhd c$ and $|h| \lhd c$ for all $h \in A$. It follows that $\lim_{i\to\infty}\|(e - \inf A_{n_i})^+\|_c = 0$ and $A_{n_i} \subset E_c$ for all $i \in I$.
Given $\varepsilon > 0$, choose $i \in \mathbb{N}$ such that $(e - \inf A_{n_i})^+ \leq \varepsilon c$ and $|Se - S(\inf A_{n_i})| \leq \varepsilon g$. In G", we then obtain the inequality

$$\inf_{\substack{h\in H_1\cap E_c \\ h\geq e-\varepsilon c}} Sh \leq \inf S(A_{n_i}) = S(\inf A_{n_i}) \leq Se + \varepsilon g,$$

where H_1 is the linear subspace of H generated by A.
Since $\varepsilon > 0$ was arbitrary, we conclude

$$Se \leq \hat{S}^c_{H_1\cap E_c}(e) = \sup_{\varepsilon>0} \inf_{\substack{h\in H_1\cap E_c \\ h\geq e-\varepsilon c}} Sh \leq Se$$

in the sup-completion of G", hence $Se = \hat{S}^c_{H_1\cap E_c}(e)$.
Similarly, using a sequence (k_n) of suprema of finite subsets of H such that $\lim_{n\to\infty} Sk_n = Se$ and $\lim_{n\to\infty}\|(e - k_n)^-\| = 0$, we can select a

countably generated linear subspace $H_2 \subset H$ and an element $d \in E_+$ such that $\check{S}^d_{H_2 \cap E_d}(e) = Se$. If H_o denotes the linear subspace of H generated by H_1 and H_2 and $a := c + d$, then $c \lhd a$, $d \lhd a$, therefore

$$\check{S}^d_{H_2 \cap E_d}(e) \leq \check{S}^d_{H_o \cap E_a}(e) \leq \check{S}^a_{H_o \cap E_a}(e) \leq Se \leq \hat{S}^a_{H_o \cap E_a}(e) \leq \hat{S}^c_{H_o \cap E_a}(e) \leq \hat{S}^c_{H_1 \cap E_c}(e).$$

Since $\check{S}^d_{H_2 \cap E_d}(e) = Se = \hat{S}^c_{H_1 \cap E_c}(e)$, we conclude that

$$\check{S}^a_{H_o \cap E_a}(e) = Se = \hat{S}^a_{H_o \cap E_a}(e).$$

The implications (ii) $\Rightarrow$ (iii), (iii) $\Rightarrow$ (iv) are evident.

(iv) $\Rightarrow$ (i): Choose $a \in E_+ \setminus \{0\}$ such that $\check{S}^a_H(e) = Se = \hat{S}^a_H(e)$. If $\hat{H}$ denotes the set of infima of all finite subsets of H, the inclusion $\{k \in E : k \geq e - \varepsilon a\} \subset \{k \in E : \|(e-k)^+\| \leq \varepsilon\|a\|\}$ yields

$$\inf_{\substack{k \in \hat{H} \\ k \geq e - \varepsilon a}} f(Sk) \geq \inf_{k \in \hat{H}_{e,\varepsilon\|a\|}} f(Sk) \text{ for all } f \in G'_+,\ \varepsilon > 0.$$

Consequently,

$$f(Se) \leq \hat{S}_H e(f) = \sup_{\varepsilon > 0} \inf_{k \in \hat{H}_{e,\varepsilon\|a\|}} f(Sk) \leq \sup_{\varepsilon > 0} \inf_{\substack{k \in \hat{H} \\ k \geq e - \varepsilon a}} f(Sk) = \sup_{\varepsilon > 0} (\inf_{\substack{k \in \hat{H} \\ k \geq e - \varepsilon a}} Sk)(f)$$

$$= \hat{S}^a_H(e)(f) = f(Se) \quad \text{for all } f \in G'_+,$$

the last infimum being formed in G". It follows that $Se = \hat{S}_H e$. Similarly, the equality $\check{S}^a_H(e) = Se$ implies $\check{S}_H e = Se$. By Theorem 6.7 we conclude $e \in \mathrm{Kor}_{\mathcal{P}_e, S}(H)$. ■

<u>6.16 Definition</u>: For each $x \in E$ we introduce the following subsets of H:

$H_x := \{h \in H : h \geq x\}$,

$H^x := \{h \in H : h \leq x\}$.

Furthermore, we set

$\hat{H}_x := \{\inf A : \emptyset \neq A \subset H_x,\ A \text{ finite}\}$,

$\check{H}^x := \{\sup A : \emptyset \neq A \subset H^x,\ A \text{ finite}\}$.

Note that $\check{H}^x$ is upward and $\hat{H}_x$ is downward directed.

<u>6.17 Corollary of Theorem 6.15</u>: If $\hat{S}_H$ is regularized and if either G" possesses the positive bounded approximation property or G is a band G", then

$$\mathrm{Kor}_{\mathcal{P}_e,S}(H) = \{e \in E : \exists a \in E_+ \forall \varepsilon > 0 \lim_{k \in \hat{H}_{e-\varepsilon a}} S(k \vee e) = Se = \lim_{k \in \check{H}^{e+\varepsilon a}} S(k \wedge e)\}.$$

<u>Proof</u>: Given $e \in \mathrm{Kor}_{\mathcal{P}_e,S}(H)$, there exists an element $a \in E_+$ such that $\check{S}^a_H(e) = Se = \hat{S}^a_H(e)$, by Theorem 6.15. In the sup-completion G''_s of G" we thus have the equality

$$\sup_{\varepsilon>0} \inf_{k \in \hat{H}_{e-\varepsilon a}} S(k \vee e) = \sup_{\varepsilon>0} \inf_{h \in H_{e-\varepsilon a}} Sh \vee Se = \hat{S}^a_H(e) \vee Se = Se,$$

Consequently, for each $f \in G'_+$:

$$f(Se) = (\sup_{\varepsilon>0} \inf_{k \in \hat{H}_{e-\varepsilon a}} S(k \vee e))(f) = \sup_{\varepsilon>0} \inf_{k \in \hat{H}_{e-\varepsilon a}} f(S(k \vee e)),$$

which shows that $f(Se) = \inf_{k \in \hat{H}_{e-\varepsilon a}} f(S(k \vee e))$ for each $\varepsilon > 0$. The decreasing net $(S(k \vee e))_{k \in \hat{H}_{e-\varepsilon a}}$ converging weakly to Se we deduce $\lim_{k \in \hat{H}_{e-\varepsilon a}} S(k \vee e) = Se$ with respect to the norm-topology (see [65], V, 4.3). Similarly, the equality $\lim_{k \in \check{H}^{e+\varepsilon a}} S(k \wedge e) = Se$, $\varepsilon > 0$, follows from $\check{S}^a_H(e) = Se$.

Conversely, assume that $\lim_{k \in \hat{H}_{e-\varepsilon a}} S(k \vee e) = Se$ for each $\varepsilon > 0$. Then $\inf_{h \in H_{e-\varepsilon a}} Sh = \inf_{k \in \hat{H}_{e-\varepsilon a}} Sk \leq Se$ in $G'' \subset G''_s$, which yields $\hat{S}^a_H(e) \leq Se$. The inequality $\hat{S}^a_H(e) \geq Se$ being always true we conclude $Se = \hat{S}^a_H(e)$. In the same way, the equality $\check{S}^a_H(e) = Se$ results from the relation $\lim_{k \in \check{H}^{e+\varepsilon a}} S(k \wedge e) = Se$, $\varepsilon > 0$. ∎

6.18 Remarks:

i) It can be shown that

$$\inf_{a\in E_+} \hat{S}^a_{H\cap E_a}(e) = \inf_{a\in E_+} \hat{S}^a_H(e) = \hat{S}_H(e) \quad \text{in } G''_s \text{ for all } e\in E.$$

This equality reflects the coincidence of the order topology (i.e. the finest locally convex topology such that all order intervals are topologically bounded) and the norm topology on E. We shall not use this relation in the sequel.

ii) The characterization of $\text{Kor}_{P_e,S}(H)$ in Theorem 6.15, ii gives us some new and surprising information that will be fundamental for the next section: Consider a single element $e\in E$ and assume that G" has the PBAP. To decide whether or not $e\in \text{Kor}_{P_e,S}(H)$ we may replace H by some countably generated linear subspace H_o. Moreover, it suffices to shrink the space E to a suitable lattice ideal E_a containing e. Finally, we may forget the original norm on E substituting it by the order unit norm $\|\cdot\|_a$ of E_a.
Thus, every efficient description of Korovkin closures in AM-spaces will almost automatically carry over to other Banach lattices and, in particular, to L^p-spaces. In the next section we shall therefore start with a practicable characterization of Korovkin closures in $C_o(X)$, X locally compact, before changing over to the L^p-case.

Final remarks to section 5:

a) There are still some annoying gaps in the theory developed in this section. Thus, e.g., it is unknown for which pairs (E,G) of Banach lattices $\hat{S}_H$ is regularized for all lattice homomorphisms $S : E \to G$ and all linear subspaces $H \subset E$. Adapted pairs of Banach lattices satisfy this condition, but we have no characterization of all adapted pairs of Banach lattices either.

b) If $\hat{S}_H$ is not regularized the conditions (ii) and (iii) of Theorem 6.7 are still sufficient for an element $e \in E$ to belong to $\mathrm{Kor}_{\mathcal{P}_e,S}(H)$ as the proof of Theorem 6.7 shows. On the other hand, Example 2.13.3 shows that the equality

$$\sup_{\substack{h \in H \\ \|(e-h)^-\| \leq \varepsilon}} (Sh \wedge Se) = Se = \inf_{\substack{h \in H \\ \|(e-h)^+\| \leq \varepsilon}} (Sh \vee Se) \quad \text{for all } \varepsilon > 0$$

is already a necessary condition for $e \in \mathrm{Kor}_{\mathcal{P}_e,S}(H)$ provided that G is Dedekind complete.

c) If $\mathcal{P}_1$ denotes the class of all nets of positive contractive operators from an L^p-space into itself, there exists a very elegant solution of the characterization problem for $\mathrm{Kor}_{\mathcal{P}_1,S}(H)$ (see [13]): $\mathrm{Kor}_{\mathcal{P}_1}(H)$ is the smallest closed vector sublattice of L^p containing H. A deduction of this result directly from Theorem 4.5 or 3.2 should be possible.

7. Korovkin theorems for the identity mapping on classical Banach lattices

Throughout this section $E(=G)$ will denote a classical Banach lattice and $S : E \to E$ will be the identity operator. As before, H is a linear subspace of E.

The first part of this section is concerned with Korovkin theorems in spaces $C_o(X)$ of continuous functions vanishing at infinity. Since many applications of Korovkin theorems deal with spaces of continuous functions there exists a long list of publications on this subject (see, e.g., [40],[63],[19],[48],[67],[6],[14],[8],[26]). A complete characterization of Korovkin closures in $C_o(X)$ efficient in the applications was first given in [9]. We shall first deduce the descriptions of Korovkin closures and Korovkin systems obtained there from the results of the preceding section. Generalizations to AM-spaces can be found in [28] basing on [75].

7.1 Notations: By $C_o(X)$ we denote the Banach lattice of all continuous real-valued functions on a fixed locally compact space X vanishing at infinity endowed with the sup-norm. For $F \in C_o(X)$ we define $\hat{f} : X \to \mathbb{R}_\infty$ and $\check{f} : X \to \mathbb{R} \cup \{-\infty\}$ by

$$\hat{f}(x) = \sup_{\varepsilon>0} \inf_{\substack{h \geq f-\varepsilon \\ h \in H}} h(x), \qquad \check{f}(x) = \inf_{\varepsilon>0} \sup_{\substack{h \leq f+\varepsilon \\ h \in H}} h(x), \qquad x \in X).$$

Thus, $\hat{f}(x) = \hat{S}_H f(\varepsilon_x)$ and $\check{f}(x) = \check{S}_H f(\varepsilon_x)$ for the identity operator $S : C_o(X) \to C_o(X)$ and the Dirac measure ε_x at x.

7.2. Lemma: The following statements are equivalent for $f \in C_o(X)$:

i) $\check{S}_H(\ell) = \ell(f) = \hat{S}_H f(\ell)$ for all $\ell \in C_o(X)'_+$,

ii) $\check{f}(x) = f(x) = \hat{f}(x)$ for all $x \in X$.

Proof: It suffices to show the implication (ii) $\Rightarrow$ (i).
Given $\varepsilon > 0$, note that $\hat{H}_{f,\varepsilon} = \{k \in H : \|(f-k)^+\| \le \varepsilon\} = \{k \in H : k \ge f-\varepsilon\}$ is downward directed. Furthermore, the set

$$K := \{\ell \in C_o(X)'_+ : \|\ell\| \le 1\}$$

is a convex $\sigma(C_o(X)', C_o(X))$-compact set with extreme boundary $\{\varepsilon_x : x \in X\} \cup \{0\}$ (see [15]). The function

$$\ell \to \inf_{k \in \hat{H}_{f,\varepsilon}} \ell(k) \qquad (\ell \in K)$$

is clearly u.s.c. with respect to $\sigma(C_o(X)', C_o(X))$. Moreover, it is affine, since $\hat{H}_{f,\varepsilon}$ is downward directed. The estimate

$$\inf_{k \in \hat{H}_{f,\varepsilon}} \varepsilon_x(k-f) = \inf_{h \ge f-\varepsilon} h(x) - f(x) \le 0$$

following from the equality $\hat{f}(x) = f(x)$ for each $x \in X$ yields

$$\inf_{k \in \hat{H}_{f,\varepsilon}} \ell(k-f) \le 0$$

for all $\ell \in K$ using Bauer's maximum principle (see [18], page 102). The same inequality then holds for all $\ell \in \bigcup_{\lambda > 0} \lambda K = C_o(X)'_+$.
We thus obtain $\inf_{k \in \hat{H}_{f,\varepsilon}} \ell(k) \le \ell(f)$ for each $\varepsilon > 0$ or, equivalently, $\hat{S}_H f(\ell) = \ell(f)$ for all $\ell \in C_o(X)'_+$.
Similarly, the equality $\check{S}_H f(\ell) = \ell(f)$, $\ell \in C_o(X)'_+$, results from the coincidence of $\check{f}$ and f. ∎

7.3 Theorem: A function $f \in C_o(X)$ is in the Korovkin closure $\mathrm{Kor}_{P_e}(H)$, iff $\check{f} = f = \hat{f}$.

This is an immediate consequence of Theorem 6.7 and Lemma 7.2. ∎

7.4 Definitions: For each $x \in X$ let $M_x(H)$ denote the set of all finite positive Radon measures μ on X satisfying the condition

$$\mu(h) = h(x) \quad \text{for all } h \in H.$$

The measures $\mu \in M_x(H)$ will be called H-representing measures for x. Adopting the usual terminology from the compact case the Choquet boundary of X with respect to H is the set

$$\partial_H(X) = \{x \in X : M_x(H) = \{\varepsilon_x\}\}.$$

Applying 2.13.3, ii to the special case $F = \mathbb{R}$ and $S = \varepsilon_x$ we immediately obtain

7.5 Lemma: If $f \in C_o(X)$ and $x \in X$, then
$\hat{f}(x) = \sup\{\mu(f) : \mu \in M_x(H)\}$ and
$\check{f}(x) = \inf\{\mu(f) : \mu \in M_x(H)\}$.
In particular, the equality $\check{f}(x) = \hat{f}(x)$ holds iff $\mu(f) = f(x)$ for all $\mu \in M_x(H)$.

7.6 Corollary: In order that H is a Korovkin space in $C_o(X)$ it is necessary and sufficient that $\partial_H(X) = X$.

7.7 Example: If, for each $x \in X$, there exists a function $h_x \in H$ such that $h_x(x) = 0$ and $h_x(y) > 0$ for all $y \in X \setminus \{x\}$, then H is a Korovkin space, provided that X contains more than one point. Indeed, every H-representing measure μ for x is carried by $\{x\}$, since $\mu(h_x) = h_x(x) = 0$, which yields $\mu = \lambda\varepsilon_x$ for some $\lambda \geq 0$. On the other hand, given $y \in X \setminus \{x\}$, the equality $h_y(x) = \mu(h_y) = (\lambda\varepsilon_x)(h_y) = \lambda h_y(x)$ shows that $\lambda = 1$. Thus, $M_x(H) = \{\varepsilon_x\}$ for each $x \in X$, i.e. $\partial_H(X) = X$.

For $X = [0,1]$ and H spanned by the three functions $x \to 1$, $x \to x$, $x \to x^2$ the original theorem of Korovkin (see [40]) is an immediate consequence of 7.7. Further examples for compact and locally compact spaces X can be found in [40],[19],[9].

Korovkin systems (with respect to the identity operator) consisting of only two elements cannot exist in a Banach lattice with algebraic dimension greater than two. This is shown in the following application of Theorem 7.3 and its corollary.

7.8 Theorem: If H is a two-dimensional linear subspace of $C_o(X)$, then $\mathrm{Kor}_{\mathcal{P}_e}(H) = H$.

Proof: Suppose that there exists a function $f \in \mathrm{Kor}_{\mathcal{P}_e}(H) \setminus H$. Given $x \in X$ and $n \in \mathbb{N}$, note that there are functions $h_n, h'_n \in H$ such that $h_n \geq f - \frac{1}{n}$, $h'_n \leq f + \frac{1}{n}$ and $h_n(x) - h'_n(x) \leq \frac{1}{n}$, since $\check{f}(x) = f(x) = \hat{f}(x)$. If $\lim_{i\to\infty}\|h_{n_i} - h'_{n_i}\| = 0$ for some subsequence $(h_{n_i} - h'_{n_i})_{i\in\mathbb{N}}$ of $(h_n - h'_n)_{n\in\mathbb{N}}$, then the estimate

$$\|h_{n_i} - f\| \leq \|(h_{n_i} - f)^+\| + \frac{1}{n_i} \leq \|(h_{n_i} - h'_{n_i})^+\| + \|h'_{n_i} - f)^+\| + \frac{1}{n_i}$$

$$\leq \|h_{n_i} - h'_{n_i}\| + \frac{2}{n_i}$$

yields $\lim_{i\to\infty} h_{n_i} = f$, i.e., $f \in H$, contradicting the assumption.

Hence $\delta := \liminf_{n\to\infty}\|h_n - h'_n\| > 0$.

We may assume that $h_n \neq h'_n$ for all $n \in \mathbb{N}$. The unit sphere in H being compact the sequence (k_n) defined by

$$k_n := \frac{h_n - h'_n}{\|h_n - h'_n\|}$$

has a convergent subsequence. Thus, without loss of generality, we may assume that $k_x := \lim_{n\to\infty} k_n$ exists in H. From the inequality

$$\|k_n^-\| \leq \frac{4}{n\cdot\delta}$$

valid for all sufficiently large $n \in \mathbb{N}$ we deduce $\|k_x^-\| = \lim_{n\to\infty}\|k_n^-\| = 0$, which shows that $k_x \in H_+$. Since also $k_n(x) \leq \frac{2}{n\delta}$ for each sufficiently large $n \in \mathbb{N}$, it follows that $k_x(x) = \lim_{n\to\infty} k_n(x) = 0$.

Observing that H is two-dimensional, the set $\{x \in X : \exists h \in H \;\; h(x) \neq 0\}$ $=: X_o$ is non-empty. Choose a fixed point $x \in X_o$. The corresponding function k_x constructed above lies in the unit sphere of H, hence there is a point $y \in X_o$ such that $k_x(y) > 0$. The associated function $k_y \in H$ is linearly independent of k_x, since $k_y(y) = 0 < k_x(y)$. Consequently, $\{k_x, k_y\}$ is a basis of H and $k_y(x) > 0$ for otherwise $x \notin X_o$. The operator $T : C_o(X) \to H$, defined by

$$Tg = \frac{g(x)}{k_y(x)} k_y + \frac{g(y)}{k_x(y)} \cdot k_x$$

is a positive projection of $C_o(X)$ onto H. Thus $f \in \mathrm{Kor}_{P_e}(H)$ yields $f = Tf \in H$ contradicting the assumption. ∎

<u>7.9 Corollary</u>: If H is a two-dimensional linear subspace of a classical Banach lattice, then $\mathrm{Kor}_{P_e}(H) = H$.

<u>Proof</u>: Given $e \in \mathrm{Kor}_{P_e}(H)$ and $\varepsilon > 0$, there exists an element $a \in E_+$ such that

$$\lim_{k\in\hat{H}_{e-\varepsilon a}} k \vee e = e = \lim_{k\in\check{H}^{e+\varepsilon a}} k \wedge e \quad \text{for all } \varepsilon > 0$$

by Corollary 6.17. Increasing a, if necessary, we may assume that $H \subset E_a$. Select a decreasing sequence (k_n) in $\hat{H}_{e-\varepsilon a}$ such that

$$\|k_n \vee e - e\| \leq \frac{1}{n} \quad \text{for each } n \in \mathbb{N}.$$

Using the equivalence of the topological convergence and the relative uniform $*$-convergence in Banach spaces we can find an element $b_\varepsilon \in E_+$ and a subsequence $(k_{n_i})_{i\in\mathbb{N}}$ of (k_n) such that $\lim_{i\to\infty}\|k_{n_i} \vee e - e\|_{b_\varepsilon} = 0$ where $\|\cdot\|_{b_\varepsilon}$ denotes the Minkowski functional of the order interval $[-b_\varepsilon, b_\varepsilon]$. Since $\hat{H}_{e-\varepsilon a}$ is downward directed, it follows that

$$b_\varepsilon \sim \lim_{k\in\hat{H}_{e-\varepsilon a}} k \vee e = e.$$

Here b_ε~lim stands for the limit with respect to the norm $\|\cdot\|_{b_\varepsilon}$ on the vector lattice ideal E_{b_ε}. Similarly, there exists $c_\varepsilon \in E_+$ such that $c_\varepsilon \sim \lim_{k\in\check{H}^{e+\varepsilon a}} k \wedge e = e.$

Given an upper bound d of the countable set $\{b_{\frac{1}{n}} : n \in \mathbb{N}\} \cup \{c_{\frac{1}{n}} : n \in \mathbb{N}\}$ with respect to the preordering $\lhd$, we obtain for $z := a + d$:

$$e = z \sim \lim_{k\in\hat{H}_{e-\varepsilon a}} k \vee e \geq z \sim \lim_{k\in\hat{H}_{e-\varepsilon z}} k \vee e \geq e \geq z\sim\lim_{k\in\check{H}^{e+\varepsilon z}} k \wedge e \leq z \sim \lim_{k\in\check{H}^{e+\varepsilon a}} k \wedge e = 0,$$

hence $z \sim \lim_{k\in\hat{H}_{e-\varepsilon z}} k \vee e = e = z \sim \lim_{k\in\check{H}^{e+\varepsilon z}} k \wedge e$, for all $\varepsilon > 0$.

By Kakutani's theorem there exists a vector lattice isomorphism $g \to \bar{g}$ of E_z onto some space $C(K)$, K compact, such that $\bar{z} = 1$. Thus

$$\inf_{\substack{h\in H \\ \bar{h}\geq\bar{e}-\varepsilon}} \bar{h}(x) \leq \bar{e}(x) \leq \sup_{\substack{h\in H \\ \bar{h}\leq\bar{e}+\varepsilon}} \bar{h}(x) \quad \text{for all } x \in K,\ \varepsilon > 0,$$

or, equivalently,

$$\sup_{\varepsilon>0} \inf_{\substack{h\in H \\ \bar{h}\geq\bar{e}-\varepsilon}} \bar{h}(x) \leq \bar{e}(x) \leq \inf_{\varepsilon>0} \sup_{\substack{h\in H \\ \bar{h}\leq\bar{e}+\varepsilon}} \bar{h}(x) \qquad (x \in K).$$

The converse inequality being always true we conclude that $e \in H$ from Theorem 7.3 and Theorem 7.8. ■

In order to test the efficiency of the results obtained in the first part of this section let us now turn to a problem that has been open for the last few years:

Since $\mathrm{Kor}_{P,S}(H)$ and $\mathrm{Kor}_{P_e,S}(H)$ only defer from each other by the equicontinuity condition one may ask whether $\overline{\mathrm{Kor}_{P,S}(H)} = \mathrm{Kor}_{P_e,S}(H)$. Indeed, if $(T_i) \in P_e$ is such that $\lim_{i\in I} T_i h = Sh$ for all $h \in H$, then we not only have

$$\lim_{i\in I} T_i e = Se \quad \text{for all } e \in \mathrm{Kor}_{P,S}(H),$$

since $(T_i) \in P$ but, moreover, the same equality holds for all $e \in \overline{\mathrm{Kor}_{P,S}(H)}$ by the equicontinuity of (T_i). Thus, at any rate, $\overline{\mathrm{Kor}_{P,S}(H)} \subset \mathrm{Kor}_{P_e,S}(H)$.

Observe now that, for <u>every</u> choice of Banach lattices E,G and every vector lattice homomorphism $S : E \to G$,

$$\mathrm{Kor}_{P,S}(H) = A_S(H) := \{e \in E : H_e \neq \emptyset,\ H^e \neq \emptyset,\ \underline{\lim}_{k\in \hat{H}_e} Sk = Se = \overline{\lim}_{k\in \check{H}^e} Sk\}$$

(see [6],[11],[20],[21]). The elements of $A_S(H)$ called <u>H-affine</u> elements of E are rather easy to determine in many cases of practical interest. Thus, $\overline{A_S(H)}$ is an easily accessible lower estimate for $\mathrm{Kor}_{P_e,S}(H)$ and H is a Korovkin space, whenever $A_S(H)$ is dense in E.

Suppose now that the equality

$$(*) \qquad \mathrm{Kor}_{P_e,S}(H) = \overline{\mathrm{Kor}_{P,S}(H)} \quad (= \overline{A_S(H)})$$

would hold. Then the major part of the results proved in the last section and the first part of this section would be of little interest since the characterization of $\mathrm{Kor}_{P_e,S}(H)$ as the topological closure of the H-affine elements were efficient, easy to check and generally true (even when $\hat{S}_H$ is not regularized).

As the following counterexample shows, however, equality (*) may fail even when $E = G = C_o(X)$ S is the identity operator and H is finite dimensional.

7.10 Counterexample: Let $g_1, g_2, g_3, g_4, g_5 : \mathbb{R} \to \mathbb{R}$ be defined as follows

$$g_1(x) = 1, \quad g_2(x) = x, \quad g_3(x) = (x+2)^3$$

$$g_4(x) = \begin{cases} (x+1)^2 & \text{for } x \in]-\infty,-1[\\ 0 & \text{if } x \in [-1,1] \\ (x-1)^2 & \text{for } x \in]1,\infty[, \end{cases}$$

$$g_5(x) = \begin{cases} |x|^3 - 3 & \text{for } x \in]-\infty,-1[\\ -2(x+2)^3 & \text{for } x \in [-1,1] \\ |x|^3 - 55 & \text{for } x \in]1,\infty[. \end{cases}$$

If $f_o(x) = \exp(-x^2)$ for all $x \in \mathbb{R}$, let H denote the linear subspace of $C_o(\mathbb{R})$ generated by $\{f_o \cdot g_1, f_o \cdot g_2, f_o \cdot g_3, f_o \cdot g_4, f_o \cdot g_5\}$. Then H has the following properties:

i) H is a Korovkin space with respect to $\mathcal{P}_e$ and the identity operator

ii) H is positively generated, i.e. $H = H_+ - H_+$,

iii) H separates the points of $\mathbb{R}$,

iv) $\mathrm{Kor}_{\mathcal{P}}(H) = A(H) \subset M := \{f \in C_o(\mathbb{R}) : \exists h \in H \ \ h|_{[-1,1]} = f|_{[-1,1]}\}$, where $A(H)$ denotes the set of all H-affine functions with respect to the identity operator.

In particular, $\overline{A(H)} \subset M \neq C_o(\mathbb{R}) = \mathrm{Kor}_{\mathcal{P}_e}(H)$.

Proof: In order to show (i) we use Corollary 7.6. Given $x_o \in \mathbb{R}$ and $\mu \in M_{x_o}(H)$ it suffices to construct a function $f \in L^1(\mu)$ satisfying

(*) $\int f d\mu = 0$, $f(x_o) = 0$ and $f(x) > 0$ for all $x \in \mathbb{R} \setminus \{x_o\}$.

Indeed, it then follows that $\{x_o\}$ is the support of μ, i.e. $\mu = \lambda \varepsilon_{x_o}$ for some $\lambda \geq 0$, and $\lambda = 1$ since $\lambda f_o(x_o) = \int f_o g_1 d\mu = (f_o g_1)(x_o) = f_o(x_o)$. If $x_o \in]-\infty,-1[$, let $g_{x_o} := g_4 - 2(x_o+1)g_2 + (x_o^2-1)g_1$. Then g_{x_o} is clearly convex on $\mathbb{R}$ and

$g_{x_o}(x) = (x+1)^2 - 2(x_o+1)x + (x_o^2-1) = (x-x_o)^2$ for all $x \in]-\infty,-1[$.

Hence, the function $f := f_o \cdot g_{x_o}$ satisfies (*).

Similarly, if $x_o \in]1,\infty[$, the function $f_o \cdot g'_{x_o}$ satisfies condition (*) for $g'_{x_o} := g_4 - 2(x_o-1)g_2 + (x_o^2-1)g_1$.

If $x_o \in [-1,1]$, we put

$$g''_{x_o} := g_3 - 3(x_o+2)^2 g_2 + 2(x_o+2)^2(x_o-1)g_1.$$

Note that $g''_{x_o}(x_o) = (x_o+2)^3 - 3(x_o+2)^2(x_o+2) + 2(x_o+2)^3 = 0$.

Computing the first and second derivative $Dg''_{x_o}(x_o)$ and $D^2g''_{x_o}(x_o)$, respectively, at x_o we obtain

$$Dg''_{x_o}(x_o) = 0 \quad \text{and} \quad D^2g''_{x_o}(x_o) = 6(x_o+2) > 0.$$

Hence g''_{x_o} has a double zero and a local minimum in x_o. Finally, $x_1 := -2(x_o+3) \leq -4$ is the second zero of g''_{x_o}. Since $\sup_{n\in\mathbb{N}}(g''_{x_o} + ng_4)(x) = \infty$ for all $x \in \mathbb{R}\setminus[-1,1]$, we conclude that $g := \inf(g_1, \sup_{n\in\mathbb{N}}(g''_{x_o} + ng_4))$ is a l.s.c., bounded function on $\mathbb{R}$ such that $g(x_o) = 0$ and $g(x) > 0$ for all $x \in \mathbb{R}\setminus\{x_o\}$. As the following estimates show, the function $f := f_o \cdot g$ satisfies condition (*):

$$0 \leq \int f d\mu = \sup_{n\in\mathbb{N}} \int f_o \cdot \inf(g_1, g''_{x_o} + ng_4) d\mu$$

$$\leq \sup_{n\in\mathbb{N}} \int f_o \cdot (g''_{x_o} + ng_4) d\mu = \sup_{n\in\mathbb{N}} (f_o \cdot (g''_{x_o} + ng_4))(x_o) = 0.$$

Thus statement (i) is a consequence of Corollary 7.6.

To prove (ii) note that the set $\{x \in \mathbb{R} : 2g_5(x) \leq g_k(x)\}$ is compact for $k = 1,\ldots,5$. Consequently, for n sufficiently large,

$$2g_5 + n \cdot g_1 \geq g_k, \qquad k = 1,2,3,4,5.$$

It follows that H is positively generated.

Since H clearly separates the points of $\mathbb{R}$, it remains to prove (iv).

The positive linear form $h \to \lim_{x\to\infty} \frac{h(x)}{|x|^3 \cdot f_o(x)}$ defined on H has a positive linear extension φ_∞ to the space

$$\widetilde{H} := \{f \in C_o(X) : \exists h_1, h_2 \in H \quad h_1 \leq f \leq h_2\}.$$

Similarly, we can find a positive linear form $\varphi_{-\infty}$ on $\widetilde{H}$ such that $\varphi_{-\infty}(h) = \lim_{x\to-\infty} \frac{h(x)}{|x|^3 \cdot f_o(x)}$ for all $h \in H$. Given $x \in [-1,1]$, let τ_x denote the positive linear form

$$\tau_x(f) = \alpha_x^{(-\infty)} \varphi_{-\infty}(f) + \alpha_x^{(-1)} f(-1) + \alpha_x^{(1)} f(1) + \alpha_x^{(\infty)} \varphi_\infty(f)$$

on $\widetilde{H}$, where

$$\alpha_x^{(\infty)} := \frac{f_o(x)}{4} \left((1-x) + 27(1+x) - 2(x+2)^3\right),$$

$$\alpha_x^{(-\infty)} := 3\alpha_x^{(\infty)}, \qquad \alpha_x^{(-1)} := \frac{(1-x)\cdot f_o(x)}{2e}, \qquad \alpha_x^{(1)} := \frac{(1+x) f_o(x)}{2e}$$

($e := \exp(1)$).
It is easy to check that $\tau_x(h_i) = h_i(x)$ for $i = 1,\ldots,5$, $h_i := f_o \cdot g_i$. Furthermore, if $f \in A(H)$ and $h \in H^f$, $h' \in H_f$, then $h(x) = \tau_x(h) \leq \tau_x(f) \leq \tau_x(h') = h'(x)$. Consequently,

$$\sup_{h\in H^f} h(x) \leq \tau_x(f) \leq \inf_{h\in H_f} h(x),$$

which yields $\tau_x(f) = f(x)$, since f is H-affine. Note that each of the functions $x \to \tau_x^{(-\infty)}$, $x \to \tau_x^{(-1)}$, $x \to \tau_x^{(1)}$, $x \to \tau_x^{(\infty)}$ defined on $[-1,1]$ is the restriction to $[-1,1]$ of some function in H. Hence the same holds for $x \to \tau_x(f) = f(x)$ $(x \in [-1,1])$. This proves that $f \in M$. ∎

<u>7.11 Remark</u>: Since infinite dimensional AM-spaces are never KB-spaces (i.e. bands in their biduals), Theorem 6.7 does not provide a characterization of $\mathrm{Kor}_{P_o,S}(H)$ in $C_o(X)$. The description of the "stationary" Korovkin closure $\mathrm{Kor}_{P_o,S}(H)$ is, in fact, rather complicated even if X is compact, (i.e. $C_o(X) = C(X)$), S is the identity mapping and H con-

tains a strictly positive function. The interested reader should consult [57] for a detailed discussion.

Once a characterization of Korovkin closures in $C_o(X)$ has been satisfactorily settled we are now able to attack the L^p-case. We restrict our attention to <u>σ-finite measure spaces</u> $(\Omega, \mathcal{A}, \mu)$. Since $G = L^p(\mu)$ is reflexive for $p \in]1,\infty[$, the sup-completion of G" is isomorphic to the imbedding cone C_p of $L^p(\mu)$ introduced in Example 1.2,d. Given an arbitrary Banach lattice E, a vector lattice homomorphism $S : E \to G$ and elements $e \in E$, $a \in E_+$, we may interpret $\hat{S}_H^a(e)$ as an equivalence class of $\mathcal{A}$-measurable functions coinciding μ-a.e. on Ω. We shall use this interpretation without explicit reference.

For $p = 1$ $G = L^p(\mu) = L^1(\mu)$ is no longer reflexive in general but G has at least order continuous norm or, equivalently, G is a vector lattice ideal in G" (see [66], Ch. II, 5.10). The following theorem states that under this condition the sup-completion G_s of G is also isomorphic to a solid subset of G''_s. As a consequence we may consider $\hat{S}_H^a(e)$ as an equivalence class of $\mathcal{A}$-measurable μ-a.e. coinciding functions on Ω in this case, too. In other words, $\hat{S}_H^a(e)$ can be identified with an element of the embedding cone C_1 introduced in Example 1.2,d.

<u>7.12 Theorem</u>: Let F be a vector lattice ideal of a Dedekind complete vector lattice G. Then there exists a lattice cone monomorphism C from the sup-completion F_s of F into the sup-completion G_s of G such that $V(F_s)$ is a solid subset of G_s. In particular,

$V(\sup A) = \sup V(A)$ for every subset $A \subset F_s$.

$V(\inf A) = \inf V(A)$ for every $A \subset F_s$ bounded from below.

<u>Proof</u>: For each $x \in F_s$, set

$$V(x) := \sup\{f \in F : f \leq x\} ,$$

where the supremum is formed <u>in G_s</u>. Since the Riesz decomposition property

$\{f \in F : f \leq x+y\} = \{f_1 \in F : f_1 \leq x\} + \{f_2 \in F : f_2 \leq y\} \qquad (x,y \in F_s)$

obviously holds in F_s, V is additive. The positive homogenity of V is evident. Let $x,y \in F_s$ be such that $V(x) \leq V(y)$. If $f \in F$ satisfies $f \leq x$, then $f \leq V(y)$ in G_s, hence

$\sup\{f' \wedge f : f' \in F,\ f' \leq y\} = V(y) \wedge f = f$ in G_s, which yields

$\sup\{f' \wedge f : f' \in F,\ f' \leq y\} = f$ <u>in G</u>.

F being a vector lattice ideal in G it follows that

$\sup\{f' \wedge f : f' \in F,\ f' \leq y\} = f$ <u>in F</u>,

and, consequently, $y \wedge f = f$ <u>in F_s</u>. Since $f \leq x$ was chosen arbitrary, we obtain in F_s

$$x = \sup\{f \in F : f \leq x\} = \sup\{y \wedge f : f \in F,\ f \leq x\} = y \wedge x,$$

or, equivalently, $x \leq y$. In particular, V is injective.

From the equalities

$$\{f \in F : f \leq x \wedge y\} = \{f_1 \wedge f_2 : f_1, f_2 \in F,\ f_1 \leq x,\ f_2 \leq y\} \text{ and}$$

$$\{f \in F : f \leq x \vee y\} = \{f_1 \vee f_2 : f_1, f_2 \in F,\ f_1 \leq x,\ f_2 \leq y\}$$

valid for each $x,y \in F_s$ we deduce that $V(x \wedge y) = V(x) \wedge V(y)$ and $V(x \vee y) = V(x) \vee V(y)$. Therefore, V is a lattice cone monomorphism and $V(F_s)$ is a vector sublattice of G_s.

It remains to prove that $z \in V(F_s)$ for every choice of $x,y \in F_s$, $z \in G_s$ satisfying $V(x) \leq z \leq V(y)$.

Selecting a fixed element $f' \in F$, $f' \leq x$, we obtain for each $f \in F$, $f \leq y$: $z \wedge f \in G$ and $f' \wedge f \leq V(x) \wedge f \leq z \wedge f \leq f$. F being a vector lattice ideal in G we conclude $z \wedge f \in F$. Moreover, we claim that $z = V(u)$ for $u := \sup\{z \wedge f : f \in F,\ f \leq y\}$ formed <u>in F_s</u>. Note first that $z \leq V(u)$ because of the equality $z = z \wedge V(y) = \sup\{z \wedge f : f \in F,\ f \leq y\}$ valid in <u>G_s</u>. To prove the converse inequality consider an element $f'' \in F$ such that $f'' \leq u$. Then $f'' = f'' \wedge u = \sup\{f'' \wedge (z \wedge f) : f \in F,\ f \leq y\}$ in F.

Since F is a vector lattice ideal in G, the same equality holds in G and therefore in G_s. It follows that $f'' \leq z$. By the definition of $V(u)$ we thus obtain $V(u) \leq z$. ∎

In order to cut short the technical framework let us assume for the rest of this section that H has a countable algebraic basis. Note that, by Theorem 6.15,ii, this assumption does not restrict the generality of the results dealing with individual elements of $\mathrm{Kor}_{P_e}(H)$. Moreover, we shall only consider the Radon measure case, i.e. we assume that Ω is a locally compact σ-compact space X and μ is a positive Radon measure on X. Finally, recall that $p \in [1,\infty[$, $E = G = L^p(\mu)$, and $S : E \to E$ is the identity mapping throughout this section.

In order to obtain efficient characterizations of Korovkin closures and Korovkin systems in L^p-spaces it is necessary to analyse functions in $\mathcal{L}^p(\mu)$ rather than the corresponding equivalence classes in $L^p(\mu)$. Slightly generalizing the notion previously used for Banach lattices, we denote by $\mathcal{L}_a$ the vector lattice ideal

$$\mathcal{L}_a := \{ f \in \mathcal{L}^p(\mu) : |f| \leq \alpha a \quad \text{for some } \alpha > 0\},$$

whenever $a : X \to \mathbb{R}_\infty$ is a non-negative numerical function (not necessarily $a \in \mathcal{L}^p(\mu)_+$). Since the measure μ will not vary in the rest of this section we shall simply write $\mathcal{L}^p$ instead of $\mathcal{L}^p(\mu)$. Moreover, we shall choose a fixed linear subspace $H \subset \mathcal{L}^p$ with countable algebraic basis such that each equivalence class $h \in H$ has a corresponding representative $h \in H$. For a μ-measurable numerical function $g : X \to \mathbb{R} \cup \{-\infty,+\infty\}$ we set

$$H_g := \{h \in H : h \geq g \quad \mu\text{-a.e.}\} \text{ and } H^g := \{h \in H : h \leq g \quad \mu\text{-a.e.}\}.$$

Note that $H \subset \mathcal{L}_a$ for some p^{th} power μ-integrable numerical function $a : x \to]0,\infty]$, since H has a countable algebraic basis. Given such a function a and $f \in \mathcal{L}^p$ let

$$\hat{f}^a(x) := \sup_{\varepsilon>0} \inf_{h\in H_{f-\varepsilon a}} h(x) \quad \text{and} \quad \check{f}^a(x) := \inf_{\varepsilon>0} \sup_{h\in H^{f+\varepsilon a}} h(x)$$

for each $x \in X$, where $\sup \emptyset = -\infty$ and $\inf \emptyset = \infty$.

<u>7.13 Lemma</u>: Let $f \in \mathcal{L}^p$ and $a : X \to]0,\infty]$ be a pth power μ-integrable function such that $f \in \mathcal{L}_a$ and $H \subset \mathcal{L}_a$. Then $\hat{f}^a$ and $\check{f}^a$ are μ-measurable numerical functions on X satisfying $\check{f}^a \leq f \leq \hat{f}^a$ μ-a.e. Moreover, $\hat{f}^a$ is a representative for the equivalence class $\hat{S}^a_H(f)$, where $S : L^p(\mu) \to L^p(\mu)$ is the identity operator and a,f are the equivalence classes of a, f, respectively. Correspondingly, $\check{f}^a$ is a member of the equivalence class $\check{S}^a_H(f)$ in the inf-completion of $L^p(\mu)$.

<u>Proof</u>: Let $\|\cdot\|_a$ denote the Minkowski functional of the order interval $[-a,a]$ in $\mathcal{L}^p$. Then $\|\cdot\|_a$ is a seminorm on $\mathcal{L}_a$ and $\|g\|_a = 0 \Leftrightarrow g(x) = 0$ for all $x \in X_a := \{y \in X : a(y) < \infty\}$ whenever $g \in \mathcal{L}_a$. Let H_1 be a countable $\|\cdot\|_a$-dense subset of H. Since, for every $x \in X_a$ and $\varepsilon > 0$,

$$\inf_{h\in H_{f-\varepsilon a}} h(x) \geq \inf_{h\in H_1\cap H_{f-2\varepsilon a}} h(x) \geq \inf_{h\in H_{f-2\varepsilon a}} h(x) ,$$

it follows that $\hat{f}^a(x) = \sup_{n\in\mathbb{N}} \inf_{h\in H_1\cap H_{f-(2/n)a}} h(x)$, which yields the μ-measurability of $\hat{f}^a$. Passing to the corresponding equivalence classes in the sup-completion C_p of $L^p(\mu)$ (see Example 1.2,d) the same argument shows that $\hat{f}^a$ is a representative for $\hat{S}^a_H(f)$. Moreover, the inequality $\hat{f}^a \geq f - \varepsilon a$ valid μ-a.e. for all $\varepsilon > 0$, implies $\hat{f}^a \geq f$ μ-a.e. Since $-\widehat{(-f)}^a = \check{f}^a$, this completes the proof. ∎

<u>7.14 Corollary</u>: Given $f \in \mathcal{L}^p$, the following statements are equivalent:

i) $f \in \mathrm{Kor}_{P_e}(H)$, where f denotes the equivalence class of f.

ii) $\check{f}^a = f = \hat{f}^a$ μ-a.e. on X for some p^{th} power μ-integrable function $a : X \to]0,\infty]$ such that $f \in \mathcal{L}_a$ and $H \subset \mathcal{L}_a$.

Proof: This is an immediate consequence of Theorem 6.15 and Lemma 7.13. ■

7.15 Corollary: If $a, b : X \to]0,\infty]$ are p^{th} power μ-integrable numerical functions such that $\mathfrak{f} \in \mathcal{L}_a$, $H \subset \mathcal{L}_a$, $b \geq a$, and if $\check{\mathfrak{f}}^a = \mathfrak{f} = \hat{\mathfrak{f}}^a$ μ-a.e., then $\check{\mathfrak{f}}^b = \mathfrak{f} = \hat{\mathfrak{f}}^b$ μ-a.e. .

7.16 Lemma: Let $a : X \to]0,\infty]$ be a pth power μ-integrable function and let $(K_n)_{n\in\mathbb{N}}$ be a sequence of pairwise disjoint, compact subsets of X such that $a|_{K_n}$ is real-valued and continuous for each $n \in \mathbb{N}$ and $\mu(X \setminus \bigcup_{n\in\mathbb{N}} K_n) = 0$. Then there exists a pth power μ-integrable function $b : X \to]0,\infty]$ satisfying the following conditions:

i) $a \leq b$ on X,

ii) $b|_{K_n}$ is real-valued and continuous for each $n \in \mathbb{N}$,

iii) for every $\varepsilon > 0$ there is an $n_\varepsilon \in \mathbb{N}$ such that

$$A_\varepsilon := \{x \in X : a(x) > \varepsilon b(x)\} \subset \bigcup_{n=1}^{n_\varepsilon} K_n .$$

Proof: If A is a μ-measurable subset of X, let 1_A denote the characteristic function of A. Since $(a \cdot 1_{X \setminus \bigcup_{n=1}^m K_n})_{m\in\mathbb{N}}$ is a decreasing sequence of pth power μ-integrable functions satisfying $\inf_{m\in\mathbb{N}} a \cdot 1_{X\setminus\bigcup_{n=1}^m K_n} = 0$ μ-a.e., there is, for each $\ell \in \mathbb{N}$, an $m_\ell \in \mathbb{N}$ such that

$$\ell \cdot \int_{X\setminus\bigcup_{n=1}^{m_\ell} K_n} a^p \, d\mu < \frac{1}{2^\ell} .$$

If we set $B_0 := \bigcup_{n=1}^{m_1} K_n$ and $B_\ell := \bigcup_{n=m_\ell+1}^{m_{\ell+1}} K_n$ $(\ell \in \mathbb{N})$, we can define $b : X \to]0,\infty]$ as follows:

$$b(x)=\begin{cases} a(x), & \text{if } x \in B_0, \\ \sqrt[p]{\ell}\cdot a(x), & \text{if } x \in B_\ell,\ \ell \in \mathbb{N}, \\ \infty, & \text{if } x \in X \setminus \bigcup_{\ell\in\mathbb{N}} B_\ell = X \setminus \bigcup_{n\in\mathbb{N}} K_n . \end{cases}$$

b is p^{th} power μ-integrable, since

$$\int b^p \, d\mu = \sum_{\ell\in\mathbb{N}} \int_{B_\ell} \ell\cdot \alpha^p \, d\mu + \int_{B_0} a^p \, d\mu$$

$$\leq \int a^p \, d\mu + \sum_{\ell=1}^{\infty} \frac{1}{2^\ell} < \infty.$$

Clearly, the function b satisfies conditions (i) and (ii). In order to show (iii), let $\varepsilon > 0$ be given, and choose $\ell \in \mathbb{N}$ such that $\varepsilon\cdot\sqrt[p]{\ell} > 1$. For all $x \in \bigcup_{m\geq\ell} B_m \cup (X \setminus \bigcup_{n\in\mathbb{N}} K_n)$ we deduce $\varepsilon b(x) \geq \varepsilon\cdot\sqrt[p]{\ell}\cdot a(x) \geq a(x)$. Hence $A \subset \bigcup_{m=0}^{\ell-1} B_m = \bigcup_{n=1}^{m_\ell} K_n$, which completes the proof. ■

<u>7.17 Definition</u>: Let $a : X \to\ ]0,\infty]$ be a pth power μ-integrable, numerical function. Given $x \in X$, $P_x^a(H)$ denotes the set of all positive linear functionals φ on $\mathcal{L}_a$ such that $\varphi(h) = h(x)$ for all $h \in H \cap \mathcal{L}_a$ and φ is bounded on $\{g \in \mathcal{L}_a : |g| \leq a\}$.

Note that a may attain the value ∞. Hence, in general, $a \notin \mathcal{L}_a$, thus $\varphi(a)$ is not defined. As a substitute, we need the boundedness of φ on $\{g \in \mathcal{L}_a : |g| \leq a\}$.

<u>7.18 Theorem</u>: Given $f \in \mathcal{L}^p$ and a pth power μ-integrable function $a : X \to\ ]0,\infty]$ such that $f \in \mathcal{L}_a$ and $H \subset \mathcal{L}_a$, there exists a pth power μ-integrable function $b : X \to\ ]0,\infty]$ such that the following conditions hold:

i) $a \leq b$ on X.

ii) for every $\varepsilon > 0$ the set $A_\varepsilon := \{x \in X : a(x) > \varepsilon b(x)\}$ is relatively compact,

iii) if $g \in H + \mathbb{R}f$ or $g = b$, then $g|_{\bar{A}_\varepsilon}$ is real-valued and continuous for each $\varepsilon > 0$,

iv) for every $\varepsilon > 0$ $\mu|_{A_\varepsilon}$ has support A_ε,

v) $\hat{f}^b(x) = \sup\{\varphi(f) : \varphi \in P_x^b(H)\}$ and $\check{f}^b(x) = \inf\{\varphi(f) : \varphi \in P_x^b(H)\}$ for all $x \in X$ satisfying $b(x) < \infty$.

Proof: Since H has a countable algebraic basis, an obvious modification of Lusin's theorem shows that there is a sequence (K_n') of disjoint, compact subsets of X such that $a|_{K_n'}$, $f|_{K_n'}$ and $h|_{K_n'}$ are continuous for all $h \in H$, $n \in \mathbb{N}$, and $\mu(X \setminus \bigcup_{n\in\mathbb{N}} K_n') = 0$. If, for each $n \in \mathbb{N}$, K_n is the support of the measure $\mu|_{K_n'}$ induced by μ on K_n', then K_n is compact, $\mu(K_n' \setminus K_n) = 0$ and $\mu(O) \neq 0$ for every subset $O \subset K_n$ which is open relative to K_n. Furthermore, the set

$$N := X \setminus \bigcup_{n\in\mathbb{N}} K_n = (X \setminus \bigcup_{n\in\mathbb{N}} K_n') \cup \bigcup_{n\in\mathbb{N}} (K_n' \setminus K_n)$$

is μ-negligible.

By Lemma 7.16 there exists a pth power μ-integrable function $b : X \to]0,\infty]$ satisfying the conditions (i) - (iii) of Lemma 7.16. Since, for every $\varepsilon > 0$, there exists a natural number n_ε such that $A_\varepsilon \subset \bigcup_{n=1}^{n_\varepsilon} K_n$, and since A_ε is relatively open in $\bigcup_{n=1}^{n_\varepsilon} K_n$, it remains to prove (v).

If $c : X \to \mathbb{R} \cup \{-\infty\}$ is p^{th} power μ-integrable, has continuous restrictions to A_ε for each $\varepsilon > 0$ and satisfies $c(x) = -\infty$ whenever $b(x) = \infty$ $(x \in X)$, we claim that $h(x) \geq c(x)$ for all $x \in X$, $h \in H_c$. Since this inequality clearly holds, when $c(x) = -\infty$, we may assume that $b(x) < \infty$. Then $x \in A_\varepsilon$ for some $\varepsilon > 0$ and the set $\{y \in A_\varepsilon : h(y) < c(y)\}$ is empty by condition (iv), being a $\mu|_{A_\varepsilon}$-negligible subset of A_ε, which is open in A_ε. In particular, for $c = f - \delta b$ we obtain $h \geq f - \delta b$ on X for each

$\delta > 0$ and $h \in H_{f-\delta b}$. Hence, if $x \in X$ is such that $b(x) < \infty$, $\varphi \in P_x^b(H)$ and $m := \sup\{\varphi(g) : g \in \mathcal{L}_b,\ g \leq b\}$, then

$$\varphi(f) - h(x) = \varphi(f-h) \leq \delta m \quad \text{for each } h \in H_{f-\delta b}\,.$$

Hence $\varphi(f) \leq \delta m + \inf_{h \in H_{f-\delta b}} h(x)$. Since $\delta > 0$ was arbitrary, we conclude that $\varphi(f) \leq \hat{f}^b(x)$. Thus, $\sup\{\varphi(f) : \varphi \in P_x^b(H)\} \leq \hat{f}^b(x)$.
Conversely, consider the numerical functional $g \to \hat{g}^b(x)$ on $\mathcal{L}_b$. Given $g \in \mathcal{L}_b$, there exists $r > 0$ such that $g \geq -rb$. Consequently, $h \in H_{-(r+1)b}$ for each $h \in H_{g-b}$. By the preceding remark, we therefore obtain $h \geq -(r+1)b$ on X, which yields

$$\hat{g}^b(x) \geq \inf_{h \in H_{g-b}} h(x) \geq -(r+1)b(x) > -\infty\,.$$

We claim that, moreover, $g \to \hat{g}^b(x)$ is sublinear and lower semicontinuous on $\mathcal{L}_b$ with respect to the order unit seminorm $\|\cdot\|_b$. If $g_1, g_2 \in \mathcal{L}_b$ and $\varepsilon > 0$, then

$$H_{g_1+g_2-\varepsilon b} \supset H_{g_1-(\varepsilon/2)b} + H_{g_2-(\varepsilon/2)b}\,,$$

which implies $\inf_{h \in H_{g_1+g_2-\varepsilon b}} h(x) \leq \inf_{h \in H_{g_1-(\varepsilon/2)b}} h(x) +$
$+ \inf_{h \in H_{g_2-(\varepsilon/2)b}} h(x)$. Since $\varepsilon > 0$ was arbitrary, we deduce that

$\widehat{g_1+g_2}^b(x) \leq \hat{g}_1^b(x) + \hat{g}_2^b(x)$. Clearly, $g \to \hat{g}^b(x)$ is positively homogeneous, if we set $0 \cdot \infty = 0$. In order to prove the lower semicontinuity of $g \to \hat{g}^b(x)$, fix $g_o \in \mathcal{L}_b$ and $\alpha < \hat{g}_o^b(x)$. Then there exists an $\varepsilon > 0$ such that $\inf_{h \in H_{g_o-\varepsilon b}} h(x) > \alpha$. For every $g \in \mathcal{L}_b$ satisfying $\|g - g_o\|_b < \varepsilon/2$ we deduce $\hat{g}^b(x) \geq \inf_{h \in H_{g-(\varepsilon/2)b}} h(x) > \alpha$ from the inclusion $H_{g-(\varepsilon/2)b} \subset H_{g_o-\varepsilon b}$.
Given $\beta < \hat{f}^b(x)$, $\beta \in \mathbb{R}$, there exists a $\|\cdot\|_b$-continuous linear form φ on $\mathcal{L}_b$ such that $\varphi(g) \leq \hat{g}^b(x)$ for all $g \in \mathcal{L}_b$ and $\varphi(f) > \beta$, by Theorem 2.9. Since $\varphi(g) \leq \hat{g}^b(x) \leq 0$ whenever $g \leq 0$, φ is positive. Finally, for each $h \in H$, $\varphi(h) \leq \hat{h}^b(x) = h(x)$, which implies that $\varphi \in P_x^b(H)$, since H

is a linear subspace of $\mathcal{L}_b$. Thus, $\hat{f}^b(x) = \sup\{\varphi(f) : f \in P_x^b(H)\}$. Similarly, $\check{f}^b(x) = \inf\{\varphi(f) : \varphi \in P_x^b(H)\}$. ∎

As a final step we now replace the positive linear forms in $P_x^b(H)$ by suitable H-representing measures refining the techniques used in $C_0(X)$. For a pth power μ-integrable function $a : X \to]0,\infty]$ we denote by $\{a = \infty\}$ and $\{a < \infty\}$ the sets $\{x \in X : a(x) = \infty\}$ and $\{x \in X : a(x) < \infty\}$, respectively.

<u>7.19 Definition</u>: Given a point $x \in X$ and a pth power μ-integrable function $a : X \to]0,\infty]$ such that $H \subset \mathcal{L}_a$, we denote by $M_x^a(H)$ the set of all positive Radon measures η on X integrating a and all functions in H and satisfying the equality

$$\int h \, d\eta = h(x) \quad \text{for all } h \in H .$$

The measures in $M_x^a(H)$ are called <u>H-representing measures for x determined by a</u>.

<u>7.20 Theorem</u>: If $f \in \mathcal{L}^p$, the following are equivalent:

i) There exists a pth power μ-integrable function $a : X \to]0,\infty]$ such that f is η-integrable and

$$\int f \, d\eta = f(x) \quad \text{for all } \eta \in M_x^a(H) \text{ and each } x \in \{a < \infty\}.$$

ii) The equivalence class $f \in L^p(\mu)$ belonging to f is a member of $\mathrm{Kor}_{P_e}(H)$.

<u>Proof</u>: (i) ⇒ (ii): Let $a : X \to]0,\infty]$ be a pth power μ-integrable function satisfying condition (i). Replacing a by $a + c$ for some strictly positive continuous $\mathcal{L}^p$-function c defined on the σ-compact space X we may assume that $\mathcal{L}_a$ contains the space $K(X)$ of all continuous functions on X with compact support. Using Theorem 7.18 we obtain a

p^{th} power μ-integrable function $b : X \to]0,\infty]$ satisfying the conditions (i) - (v) of 7.18. If $x_o \in \{b < \infty\}$, we claim that, for each positive linear form $\varphi \in P^b_{x_o}(H)$, there exists an H-representing measure $\eta \in M^b_{x_o}(H)$ such that f is η-integrable and $\varphi(f) = \int f d\mu = f(x_o)$. To prove this let $C_{o,b}$ denote the set of all functions $g \in \mathcal{L}^p$ such that $g|_{\bar{A}_\varepsilon}$ is continuous and $\{x \in X : |g(x)| > \varepsilon b(x)\}$ is contained in some A_δ, $\delta > 0$ for each $\varepsilon > 0$, where $A_\varepsilon := \{x \in X : a(x) > \varepsilon b(x)\}$. Given $g \in C_{o,b}$, $g \geq 0$, with support contained in some $\bar{A}_\varepsilon$, $\varepsilon > 0$, we obtain

$$\int g d\eta = \inf\{\eta(k) : k \in K(X),\ g \leq k\} \geq \varphi(g)$$

for the positive Radon measure $\eta := \varphi|_{K(X)}$, since g is upper semi-continuous. Fixing a real number $\varepsilon > 0$ and a function $k \in K(X)$, $k \geq g$, we can choose a decreasing sequence (k_n) in $K(X)$, $k \geq k_n \geq g$, with infimum $g = \inf_{n \in \mathbb{N}} k_n$. Select $\delta > 0$, such that $\{x \in X : k(x) > \varepsilon b(x)\} \subset A_\delta$. Then $((k_n - \varepsilon b)^+|_{\bar{A}_\delta})_{n \in \mathbb{N}}$ is a decreasing sequence of continuous functions with infimum $(g - \varepsilon b)^+|_{\bar{A}_\delta}$.

By Dini's theorem, there exists $n \in \mathbb{N}$ such that $k_n - \varepsilon b \leq g - \varepsilon b + \varepsilon a \leq g$, since $a|_{\bar{A}_\delta}$ is continuous and strictly positive. Consequently, $\varphi(k_n - g) \leq \varepsilon \cdot m$, where $m := \sup\{\varphi(g') : g' \in \mathcal{L}_b,\ |g'| \leq b\}$. It follows that

$$\int g d\eta = \inf_{n \in \mathbb{N}} \eta(k_n) \leq \varphi(g) + \varepsilon m.$$

Since $\varepsilon > 0$ was arbitrary, we conclude that $\varphi(g) = \int g d\eta$.

Given $g \in C_{o,b}$, $g \geq 0$, such that $g(x) = 0$ whenever $b(x) = \infty$, let

$$g'(x) = \begin{cases} \sqrt{g(x) \cdot b(x)} & , \text{ if } b(x) < \infty \\ 0 & , \text{ if } b(x) = \infty. \end{cases}$$

Then the support of $g_n := (g - \frac{1}{n} g')^+$ is contained in the closure of $\{x \in X : g(x) > \frac{1}{n^2} b(x)\}$, which is a subset of some $\bar{A}_\delta$, $\delta > 0$. Thus, by the preceding remarks, $\int g_n d\eta = \varphi(g_n)$ for each $n \in \mathbb{N}$. Moreover, since (g_n) is an increasing sequence with supremum g, we obtain

$$\int g\,d\eta = \sup_{n\in\mathbb{N}} \int g_n\,d\eta = \sup_{n\in\mathbb{N}} \varphi(g_n) \le \varphi(g).$$

Observing that $g - g_n = g \wedge \frac{1}{n} g' \le \frac{\varkappa}{n} b$, where $\varkappa \in \mathbb{R}_+$ satisfies $g \le \varkappa^2 b$, we conclude that $\varphi(g - g_n) \le \frac{\varkappa}{n}\cdot m$ for each $n \in \mathbb{N}$. Consequently, $\varphi(g) = \sup_{n\in\mathbb{N}} \varphi(g_n) = \int g\,d\mu$.

In order to show that $\int g\,d\eta = \varphi(g)$ for all $g \in C_{o,b}$ it remains to prove that $N := \{b = \infty\} = X \setminus \bigcup_{n\in\mathbb{N}} A_{1/n}$ is η-negligible. Note first, that N is η-measurable, since each set $A_{1/n}$ is relatively open in the compact set $\bar{A}_{1/n}$. Given a compact subset $K \subset N$, $X \setminus \bar{A}_\varepsilon$ is an open neighborhood of K for each $\varepsilon > 0$. Let K' be a compact neighborhood of K. Then $r := \inf_{x\in K'} a(x) > 0$. For each $k \in K(X), 0 \le k \le 1$ such that $k(x) = 1$ whenever $x \in K$ and the support of k is contained in $K' \cap (X \setminus \bar{A}_\varepsilon)$ we obtain $k(x) \le \frac{1}{r} a(x) \le \frac{\varepsilon}{r} b(x)$ for all $x \in X$. Hence $\eta(K) \le \eta(k) = \varphi(k) \le \frac{\varepsilon}{r} m$. Since $\varepsilon > 0$ was arbitrary, we conclude $\eta(K) = 0$. This proves that N is η-negligible. Thus, if $g \in C_{o,b}$ and

$$\bar{g}(x) = \begin{cases} g(x) & \text{for } x \in X \setminus N \\ 0 & \text{for } x \in N, \end{cases}$$

then $g - \bar{g} \le \varepsilon b$ for all $\varepsilon > 0$ and, consequently, $\varphi(g) = \varphi(\bar{g}) = \int \bar{g}\,d\eta = \int g\,d\eta$.

In particular, $\int f\,d\eta = \varphi(f)$ and $\int h\,d\eta = \varphi(h) = h(x_o)$ for all $h \in H$. Moreover, the functions

$$b_n(x) = \begin{cases} b(x) & \text{if } x \in \bar{A}_{1/n} \\ 0 & \text{else} \end{cases}$$

form an increasing sequence in $C_{o,b}$ with supremum $\bar{b}$ (defined like $\bar{g}$ above). Thus the estimate

$$\int b\,d\eta = \int \bar{b}\,d\eta = \sup_{n\in\mathbb{N}} \int b_n\,d\eta = \sup_{n\in\mathbb{N}} \varphi(b_n) \le m$$

shows that b is η-integrable, which yields $\eta \in M^b_{x_o}(H)$.

Since $x_o \in \{b < \infty\}$ was arbitrary, it follows from condition (i) that

$\varphi(f) = f(x)$ for all $x \in \{b < \infty\}$, $\varphi \in P_x^b(H)$. Consequently, $\check{f}^b(x) = f(x) = \hat{f}^b(x)$ for all $x \in \{b < \infty\}$ by condition (v) of Theorem 7.18. Using Corollary 7.14, (ii) follows.

(ii) $\Rightarrow$ (i): Choose p^{th} power μ-integrable functions $a, a' : X \to]0,\infty]$ such that $f \in \mathcal{L}_a$, $H \subset \mathcal{L}_a$, $\check{f}^a = f = \hat{f}^a$ μ-a.e., $a' \geq a$ on X and $\{x \in X : \check{f}^a(x) \neq \hat{f}^a(x)\} \subset \{a' = \infty\}$. By Theorem 7.18 there exists a pth power μ-integrable function $b : X \to]0,\infty]$ satisfying the conditions (i) - (v) of 7.18 with a' instead of a. For each $x \in \{b < \infty\}$ we then obtain

$$\hat{f}^b(x) \leq \hat{f}^{a'}(x) \leq \hat{f}^a(x) = f(x) = \check{f}^a(x) \leq \check{f}^{a'}(x) \leq \check{f}^b(x)$$

which yields $\hat{f}^b(x) = f(x) = \check{f}^b(x) = \varphi(f)$ for all $\varphi \in P_x^b(H)$ by condition 7.18, v. Thus the proof will be complete if we can show that, for each $\eta \in M_x^b(H)$, there exists $\varphi \in P_x^b(H)$ such that f is η-integrable and $\varphi(f) = \int f d\eta$.

Since the positive linear form $g \to \int g d\eta$, defined on $\mathcal{L}_b \cap \mathcal{L}^1(\eta)$, is bounded from above on $\{g \in \mathcal{L}_b \cap \mathcal{L}^1(\eta) : g \leq b\}$, there is a $\|\cdot\|_b$-continuous, positive extension φ of $g \to \int g d\eta$ to the space $\mathcal{L}_b$ (cf. [36], p. 82, 83). Obviously, $\varphi \in P_x^b(H)$ and $\varphi(f) = \int f d\eta$. ■

In order to decide whether the equivalence class of a given function $f \in \mathcal{L}^p$ belongs to the Korovkin closure $\mathrm{Kor}_{P_e}(H)$ of H, only the effect of the H-representing measures on f is important. In fact, the proof of Theorem 7.20 shows that there is a certain indefiniteness of the vector sublattice of $\mathcal{L}^p$ containing $H + \mathbb{R}f$, on which the H-representing measures act as positive linear functionals, since it is possible to replace the pth power μ-integrable function $a : X \to]0,\infty]$ by an arbitrary p^{th} power μ-integrable function $a' : X \to]0,\infty]$, $a' \geq a$. Moreover, condition (v) of Theorem 7.18 indicates that we may substitute $M_x^a(H)$ by $P_x^a(H)$, which means that we are dealing with "H-representing functionals" rather than H-representing measures. Indeed, there are many positive linear functionals on $\mathscr{L}_a$ that are not representable by positive Radon measures on X. (On the other hand, the proof of Theorem 7.20 demonstrates that the restrictions of positive linear forms on $\mathscr{L}_a$ to a space of type $C_{o,a}$ are representable by positive Radon measures on X).

Looking for a "natural" or minimal defining space for the H-representing measures and functionals, we may suggest to substitute the set $M_x^a(H)$ in Theorem 7.20, i, by the set of all positive linear forms φ on $H+\mathbb{R}\mathfrak{f}$ satisfying $\varphi(h) = h(x)$ for all $h \in H$. This, however, leads to a description of $\mathrm{Kor}_p(H)$ instead of $\mathrm{Kor}_{p_e}(H)$ (see [20],[21]) which is in general by far smaller than $\mathrm{Kor}_{p_e}(H)$. Yet, if H is finite-dimensional, there is an "intrinsic" characterization of $\mathrm{Kor}_{p_e}(H)$ that realizes the concept outlined above modified by μ-negligible sets. This will be shown in the rest of this section.

__7.21 Theorem__: A linear subspace $H \subset \mathscr{L}^p$ possessing a countable algebraic basis generates a Korovkin space $H = \{h \in L^p(\mu) : \exists h \in H \;\; h \in h\}$ if there exists a pth power μ-integrable function $a : X \to]0,\infty]$ such that $H \subset \mathscr{L}_a$ and $M_x^a(H) = \{\varepsilon_x\}$ for all $x \in \{a < \infty\}$. The converse is also true, provided that X is second countable.

__Proof__: Let $a : X \to]0,\infty]$ be a pth power μ-integrable function such that $M_x^a(H) = \{\varepsilon_x\}$ for all $x \in \{a < \infty\}$. We may assume w.l.o.g. that $\mathscr{L}_a$ is dense in $\mathscr{L}^p$, since X is σ-compact. By Theorem 7.20 the vector lattice ideal $L_a \subset L^p(\mu)$ corresponding to $\mathscr{L}_a$ is contained in $\mathrm{Kor}_{p_e}(H)$. L_a being dense in $L^p(\mu)$ and conclude that $\mathrm{Kor}_{p_e}(H) = L^p(\mu)$, since $\mathrm{Kor}_{p_e}(H)$ is closed.

Conversely, let X be second countable and let H be a Korovkin space in $L^p(\mu)$. Then the space $K(X)$ of all continuous real-valued functions on X with compact support possesses a countable dense subset $\mathcal{D}$ with respect to the inductive topology. For each $d \in \mathcal{D}$ let $a_d : X \to]0,\infty]$ be a pth power μ-integrable function such that $\eta(d) = d(x)$ for all $\eta \in M_x^{a_d}(H)$ and all $x \in \{a_d < \infty\}$, by Theorem 7.20. Since $\mathcal{D}$ is countable there exists a pth power μ-integrable function $a : X \to]0,\infty]$ satisfying $\mathcal{L}_{a_d} \subset \mathcal{L}_a$ for each $d \in \mathcal{D}$ and $\{a = \infty\} = \bigcup_{d \in \mathcal{D}} \{a_d = \infty\}$. Then $\eta(d) = d(x)$ for all $\eta \in M_x^a(H)$, $x \in \{a < \infty\}$ and all $d \in \mathcal{D}$. It follows that $M_x^a(H) = \{\varepsilon_x\}$ for all $x \in \{a < \infty\}$. ∎

In the applications of Korovkin theorems mostly finite-dimensional test spaces $H \subset E$, $E = C_o(X)$ or $E = L^p(\mu)$, occur. Although there are many publications on this subject [see, e.g. [40],[63],[48],[14],[30],[72]), finite Korovkin systems in L^p-spaces have not yet been satisfactorily characterized. In order to specialize the preceding results to the finite dimensional case, we need the following lemma, which is closely related to the theorems of Daniell-Stone and Carathéodory (note, however, that the functions concerned are non-continuous, in general).

7.22 Lemma: Let Y be a non-empty set, E a linear sublattice of $\mathbb{R}^Y$ and let $M \neq \{0\}$ be an r-dimensional linear subspace of E, $r \in \mathbb{N}$. For every positive linear form ℓ on E satisfying $\lim_{n\to\infty} \ell(f_n) = 0$ whenever (f_n) is a decreasing sequence in E_+ such that $\inf_{n \in \mathbb{N}} f_n(y) = 0$ for all $y \in Y$, there exist r points $y_1,\ldots,y_r \in Y$ and real numbers $a_1,\ldots,a_r \in \mathbb{R}_+$ such that $\ell(g) = \sum_{i=1}^{r} a_i g(y_i)$ for all $g \in M$.

Proof: If $r = 1$, there is nothing to prove. Using induction let us assume that $r \geq 2$ and that the assertion holds for all linear subspaces of E of dimension $< r$. The convex cone M_+^* of all positive linear forms on M is closed with respect to the unique Hausdorff linear topology on the finite dimensional algebraic dual M^* of M. For each $y \in Y$, let

$\varepsilon_y \in M_+^*$ be given by $\varepsilon_y(g) = g(y)$, $g \in M$. Then the convex hull C of $\bigcup_{y \in Y} \mathbb{R}_+ \varepsilon_y$ is dense in M_+^* by the bipolar theorem. Suppose that $\ell|_M \notin C$. Then there exists $g_o \in M$ such that $\ell(g_o) \leq 0$ and $\eta(g_o) \geq 0$ for all $\eta \in C$. In particular, $\varepsilon_y(g_o) = g_o(y) \geq 0$ for all $y \in Y$ which implies that $\ell(g_o) = 0$. Let $Y_o := \{y \in Y : g_o(y) = 0\}$ and choose an algebraic supplement M_o of $\mathbb{R}g_o$ in M. If $E_1 := \{f|_{Y_o} : f \in E\}$ and $M_1 := \{g|_{Y_o} : g \in M_o\}$, then E_1 is a linear sublattice of $\mathbb{R}^{Y_o}$ and M_1 has dimension $< r$. Moreover, for every function $f \in E$ such that $f \leq 0$ on Y_o the sequence $((f - ng_o)^+)_{n \in \mathbb{N}}$ is decreasing with pointwise infimum 0 on Y, hence

$\ell(f)^+ = \lim_{n\to\infty}(\ell(f - ng_o))^+ \leq \lim_{n\to\infty} \ell((f - ng_o)^+) = 0$. Consequently, $\ell(f) \leq 0$.

Hence, for every $f \in E$ satisfying $f = 0$ on Y_o, $\ell(f) = 0$, and ℓ induces a positive linear form ℓ_1 of E_1.

Finally, if (f_n') is a decreasing sequence of non-negative functions in E_1 with pointwise infimum 0 on Y_o, then there exists a sequence (f_n) in E such that $f_n|_{Y_o} = f_n'$ for each $n \in \mathbb{N}$. The sequence $(((\inf_{1\leq i\leq n} f_i) - ng_o)^+)_{n\in\mathbb{N}}$ is then decreasing in E with pointwise infimum 0 on Y. On the other hand, $\ell(((\inf_{1\leq i\leq n} f_i) - ng_o)^+) = \ell_1(f_n')$, since $f_n' = ((\inf_{1\leq i\leq n} f_i) - ng_o)^+$ on Y_o. Consequently, $\lim_{n\to\infty} \ell_1(f_n') = \lim_{n\to\infty} \ell_1(((\inf_{1\leq i\leq n} f_i) - ng_o)^+) = 0$. From the induction hypothesis it follows that there exist $y_1,\ldots,y_s \in Y_o$ and $a_1,\ldots,a_s \in \mathbb{R}_+$, where $s < r$, such that $\ell(g) = \ell_1(g|_{Y_o}) = \sum_{i=1}^{s} a_i g(y_i)$ for all $g \in M_o$, and hence for all $g \in M$, contradicting the assumption $\ell \notin C$. ∎

Although Lemma 7.22 is intended mainly to be applied in vector lattices of non-continuous functions, we may also derive from it the following characterization of Korovkin closures in $C_o(X)$, X locally compact, when H is finite-dimensional. Alternatively, one may use Caratheodory's theorem to prove this addendum to the results obtained in

$C_o(X)$:

7.23 Theorem: Let H be an n-dimensional linear subspace of $C_o(X)$, $n \in \mathbb{N}$. Then the following statements are equivalent for $f \in C_o(X)$:

i) $f \in \mathrm{Kor}_{P_e}(H)$.

ii) For each $x \in X$ and every choice of $n+1$ non-negative real numbers $\alpha_1, \ldots, \alpha_{n+1}$ and points $x_1, \ldots, x_{n+1} \in X$ such that

$$\sum_{i=1}^{n+1} \alpha_i h(x_i) = h(x) \quad \text{for all } h \in H$$

the same equality holds for f instead of h.

Proof: By Lemma 7.22, the restriction to $H + \mathbb{R}f$ of a H-representing measure $\eta \in M_x(H)$ is of the form

$$\eta\big|_H = \sum_{i=1}^{n+1} \alpha_i \varepsilon_{x_i}\big|_H \quad \alpha_1, \ldots, \alpha_{n+1} \in \mathbb{R}_+, \; x_1, \ldots, x_{n+1} \in X.$$

Thus, the assertion follows from Theorem 7.3 and Lemma 7.5. ∎

7.24 Corollary (see [63],[30]): For an n-dimensional linear subspace of $C_o(X)$, $n \in \mathbb{N}$, the following conditions are equivalent:

i) H is a Korovkin space in $C_o(X)$,

ii) for each $x \in X$ and every choice of $n+1$ non-negative real numbers $\alpha_1, \ldots, \alpha_{n+1}$ and points $x_1, \ldots, x_{n+1} \in X$ such that $\sum_{i=1}^{n+1} \alpha_i h(x_i) = h(x)$ for all $h \in H$, it follows that $\sum_{i=1}^{n+1} \alpha_i = 1$ and $x_i = x$ whenever $\alpha_i \neq 0$.

Returning to the $\mathcal{L}^p$-setting, $1 \leq p < \infty$, recall that we assumed X to be a σ-compact, locally compact space and that μ is a positive Radon measure on X, $\mathcal{L}^p := \mathcal{L}^p(\mu)$. We need the following

7.25 Definition: Let $H \subset \mathcal{L}^p$ be finite-dimensional. Given a μ-negligible set $N \subset X$ set

$$M_x^N(H) := \{ \sum_{i=1}^{n+1} \alpha_i \varepsilon_{x_i} : (x_i) \in (X\setminus N)^{n+1}, (\alpha_i) \in \mathbb{R}_+^{n+1}$$

$$\text{and } h(x) = \sum_{i=1}^{n+1} \alpha_i h(x_i) \text{ for all } h \in H\},$$

where n is the dimension of H and ε_{x_i} denotes the evaluation functional $\varepsilon_{x_i}(f) = f(x_i)$ $(f \in \mathcal{L}^p)$.

7.26 Theorem: If H is finite-dimensional, then, for each $f \in \mathcal{L}^p$, the following statements are equivalent:

i) There exists a μ-negligible subset $N \subset X$ such that $\varphi(f) = f(x)$ for all $x \in X\setminus N$ and all $\varphi \in M_x^N(H)$.

ii) $f \in \mathrm{Kor}_{P_e}(H)$, where $f \in L^p(\mu)$ is the equivalence class of f and $H \subset L^p(\mu)$ corresponds to H.

Proof: (i) $\Rightarrow$ (ii): Let $a : X \to]0,\infty]$ be a pth power μ-integrable function such that $H \subset \mathcal{L}_a$, $f \in \mathcal{L}_a$ and $N \subset \{a = \infty\}$. Given $x \in \{a < \infty\}$ each H-representing measure $\eta \in M_x^a(H)$ induces a positive linear form ℓ on $E_0 := \{g|_{X\setminus N} : g \in \mathcal{L}^1(\eta)\}$, since N is η-negligible. For every decreasing sequence (f_n) in E_0 with pointwise infimum 0 on $X\setminus N$ we have $\lim_{n\to\infty} \ell(f_n) = \lim_{n\to\infty} \int \tilde{f}_n \, d\eta = 0$, where

$$\tilde{f}_n(y) = \begin{cases} f_n(y) & \text{for } y \in X\setminus N \\ 0 & \text{for } y \in N. \end{cases}$$

Let n be the dimension of H. By Lemma 7.22, there exist $x_1,\dots,x_{n+1} \in X\setminus N$ and $\alpha_1,\dots,\alpha_{n+1} \in \mathbb{R}_+$ such that $\sum_{i=1}^{n+1} \alpha_i g(x_i) = \ell(g|_{X\setminus N}) = \int g \, d\eta$ for all $g \in H + \mathbb{R}f$. In particular, $\varphi := \sum_{i=1}^{n+1} \alpha_i \varepsilon_{x_i} \in M_x^N(H)$, and hence

$\int f d\eta = \varphi(f) = f(x)$ by condition (i). Thus, (ii) follows from Theorem 7.20.

(ii) $\Rightarrow$ (i): By Theorem 7.20 there exists a pth power μ-integrable function $a : X \to]0,\infty]$ such that $H \subset \mathscr{L}_a$, $f \in \mathscr{L}_a$, f is η-integrable and $\int f d\eta = f(x)$ for all $\eta \in M_x^a(H)$ and all $x \in \{a < \infty\}$. Since $\{\varphi|_{K(X)} : \varphi \in M_x^N(H)\} \subset M_x^a(H)$ for $N := \{a = \infty\}$, condition (i) follows. ∎

Using an obvious modification of these arguments we deduce from Theorem 7.21:

<u>7.27 Theorem</u>: A finite-dimensional linear subspace $H \subset \mathscr{L}^p$ generates a Korovkin space $H = \{h \in L^p(\mu) : \exists h \in H \quad h \in h\}$ if there exists a μ-negligible set $N \subset X$ such that $M_x^N(H) = \{\varepsilon_x\}$ for all $x \in X \setminus N$. The converse is also true provided that the topology on X is second countable.

Generalizing results known in ℓ^p-spaces (see [38]) we obtain the following

<u>7.28 Corollary</u>: Given $p,q \in [1,\infty[$ such that $\mathscr{L}^p(\mu) \subset \mathscr{L}^q(\mu)$, let H be a finite dimensional linear subspace of $\mathscr{L}^p(\mu)$. If H denotes the corresponding subspace of $L^p(\mu)$ and $\mathrm{Kor}^p_{P_e}(H)$, $\mathrm{Kor}^q_{P_e}(H)$ are the respective Korovkin closures of H in $L^p(\mu)$ and $L^q(\mu)$, then

$$\mathrm{Kor}^q_{P_e}(H) \cap L^p(\mu) = \mathrm{Kor}^p_{P_e}(H).$$

In particular, H is a Korovkin space in $L^p(\mu)$ iff it is a Korovkin space in $L^q(\mu)$.

The statement of Corollary 7.28 is not true for infinite dimensional linear subspaces $H \subset L^p(\mu)$, in general:

7.29 Counterexample: Let λ denote the Lebesgue measure on $X := [0,1]$, $p := 2$, $q := 1$. For a non-negative element $f \in L^2(\lambda) \setminus L^\infty(\lambda)$ set

$$H := \{h \in L^2(\lambda) : \int fhd\lambda = 0\},$$

i.e. H is the hyperplane orthogonal to f in the Hilbert space $L^2(\lambda)$. If $f' \in L^2(\lambda)$ satisfies $\int f'hd\lambda = 0$ for all $h \in H$, then $f' = \alpha f$ for some $\alpha \in \mathbb{R}$. Note that H is dense in $L^1(\lambda)$. To prove this, it suffices to show that f is an L^1-clusterpoint of H. Suppose to the contrary that f is not in the L^1-closure $\bar{H}^1$ of H. Then there is a continuous linear form ℓ on $L^1(\lambda)$ such that $\ell(h) = 0$ for all $h \in H$ and $\ell(f) \neq 0$. Since

$$\ell(k) = \int kgd\lambda \quad \text{for some } g \in L^\infty(\lambda) \text{ and each } k \in L^1(\lambda)$$

we obtain $g = \alpha f$, $\alpha \in \mathbb{R} \setminus \{0\}$. This contradicts the relation $f \notin L^\infty(\lambda)$. Consequently, $L^1(\lambda) = \bar{H}^1 \subset \mathrm{Kor}^1_{P_e}(H) \subset L^1(\lambda)$, i.e. $\mathrm{Kor}^1_{P_e}(H) = L^1(\lambda)$. On the other hand, given a finite subset $A \subset H$, $A \neq \emptyset$, the estimate

$$\int (\inf A) \cdot fd\lambda \leq \inf_{h \in A} \int hfd\lambda = 0$$

yields $\int kfd\lambda \leq 0$ for all $k \in \mathrm{Kor}^2_{P_e}(H)$, by Theorem 6.7, ii. Since $\mathrm{Kor}^2_{P_e}(H)$ is a linear subspace of $L^2(\lambda)$, it follows that

$$\int kfd\lambda = 0 \quad \text{for all } k \in \mathrm{Kor}^2_{P_e}(H),$$

which implies that $\mathrm{Kor}^2_{P_e}(H) = H \neq L^2(\lambda) = \mathrm{Kor}^1_{P_e}(H) \cap L^2(\lambda)$. ∎

7.30 Example: Given $p \in [1,\infty[$ let μ be a positive Radon measure on $\mathbb{R}^2$ such that the Euclidean norm is p^{th} power μ-integrable. Then the equivalence classes of the following five functions $h_1,\ldots,h_5$ form a Korovkin system in $L^p(\mu)$:

$$h_1(\xi,\eta) = 1, \qquad h_2(\xi,\eta) = \xi, \qquad h_3(\xi,\eta) = \eta,$$

$$h_4(\xi,\eta) = e^{-(\xi^2+\eta^2)}, \qquad h_5(\xi,\eta) = e^{-2(\xi^2+\eta^2)}.$$

Proof: Consider a point $x_o \in \mathbb{R}^2$ and $x_1,\dots,x_6 \in \mathbb{R}^2$, $\alpha_1,\dots,\alpha_6 \in \mathbb{R}_+$ such that $\sum_{i=1}^{6} \alpha_i h_j(x_i) = h_j(x_o)$ for $j = 1,\dots,5$. If we define $h_o(x) = (e^{-\|x\|^2} - e^{-\|x_o\|^2})^2$, where $\|\cdot\|$ denotes the Euclidean norm, then h_o is contained in the linear span H of $\{h_1,\dots,h_5\}$. Consequently, $\sum_{i=1}^{6} \alpha_i h_o(x_i) = h_o(x_o) = 0$, which implies that

$$x_1,\dots,x_6 \in \{x \in \mathbb{R}^2 : h_o(x) = 0\} = \{x \in \mathbb{R}^2 : \|x\| = \|x_o\|\} =: S.$$

On the other hand, if $(x,y) \to \langle x|y\rangle$ denotes the natural scalar product on $\mathbb{R}^2$ then the function $h(x) = \langle x_o | x_o - x\rangle$, $x \in \mathbb{R}^2$, is a member of H and $h(x) > 0$ for all $x \in S\setminus\{x_o\}$. Hence it follows from the equality $\sum_{i=1}^{6} \alpha_i h(x_i) = h(x_o) = 0$ that $\sum_{i=1}^{6} \alpha_i \varepsilon_{x_i} = \tau\varepsilon_{x_o}$ for some $\tau \geq 0$. Finally, from the relation $1 = h_1(x_o) = \sum_{i=1}^{6} \alpha_i h_1(x_i) = \tau$, we conclude that $\sum_{i=1}^{6} \alpha_i \varepsilon_{x_i} = \varepsilon_{x_o}$. Since $x_o \in \mathbb{R}^2$ was arbitrary, we obtain $M^{\emptyset}_{x_o}(H) = \{\varepsilon_{x_o}\}$ for each $x_o \in \mathbb{R}^2$. Therefore H is a Korovkin space in $L^p(\mu)$ by Theorem 7.27. ■

Examining Example 7.30 one might expect that a finite-dimensional linear subspace $H \subset \mathscr{L}^p(\mu)$ consisting only of continuous functions generates a Korovkin space $H \subset L^p(\mu)$ iff $M^{\emptyset}_x(H) = \{\varepsilon_x\}$ for μ-almost every $x \in X$. By Theorem 7.26 this condition is obviously sufficient, but even if X is a compact interval in $\mathbb{R}$ and μ is the Lebesgue measure on X, there are finite dimensional Korovkin spaces H in $L^p(\mu)$ such that the equality $M^{\emptyset}_x(H) = \{\varepsilon_x\}$ for μ-a.e. $x \in X$ is not satisfied. Moreover, the situation can not be improved by the assumption that H should contain the constant functions and separates the points of X, as the corollary of the following example shows.

The same example also demonstrates that a concept of determining the Korovkin subspaces of $L^p(\mu)$ which seems to go back to Dzjadyk [23] and

Krasnosel'skii - Lifšic [41] is not effective, in general. To outline the main ideas, let X be a compact space, μ a positive Radon measure on X and let H be a linear subspace of $C(X)$. Given $p \in [1,\infty[$ consider an equicontinuous net (T_i) of positive operators on $\mathcal{L}^p(\mu)$ such that $(T_i h)_{i\in I}$ converges to h with respect to the $\mathcal{L}^p$-semi-norm for each $h \in H$. The net $(T_i f)_{i\in I}$ will converge to f for each $f \in \mathcal{L}^p(\mu)$ iff the same holds for all $f \in C(X)$, since $C(X)$ is dense in $\mathcal{L}^p(\mu)$ and (T_i) is equicontinuous. Hence, if $\tilde{T}_i f$ denotes the equivalence class of $T_i f$ in $L^p(\mu)$ for each $f \in C(X)$, then $(\tilde{T}_i)_{i\in I}$ is a net of positive operators from $C(X)$ into $L^p(\mu)$ such that

$$\lim_{i\in I} \tilde{T}_i h = Sh \quad \text{in } L^p(\mu) \quad \text{for all } h \in H,$$

where $S : C(X) \to L^p(\mu)$ is the natural imbedding operator. It follows that the equality $\mathrm{Kor}_{p,S}(H) = C(X)$ implies that the linear subspace H of $L^p(\mu)$ corresponding to H is a Korovkin space in $L^p(\mu)$.
Since a function $f \in C(X)$ lies in $\mathrm{Kor}_{p,S}(H)$, iff f is H-affine with respect to S, i.e.

$$\inf_{\substack{h\in H\\ h\geq f}} h(x) = f(x) = \sup_{\substack{h\in H\\ h\leq f}} h(x) \quad \text{for } \mu\text{-a.e. } x \in X,$$

we retrieve the setting developed before Counterexample 7.10.
It can be shown (see [11]) that $A_S(H) = C(X)$ if and only if $M_x^{\emptyset}(H) = \{\varepsilon_x\}$ for μ-a.e. $x \in X$.
Furthermore, note that the set $A(H)$ of all H-affine elements with respect to the identity operator of $L^p(\mu)$ contains the equivalence classes of functions in $A_S(H)$. Indeed, the equivalence class f of a function $f \in \mathcal{L}^p(\mu)$ is in $A(H)$, iff

$$\inf_{h\in H_f} h = \lim_{k\in \hat{H}_f} k = f = \lim_{k\in \check{H}^f} k = \sup_{h\in H^f} h \quad \text{in } L^p(\mu),$$

or, equivalently,

$$\inf\{h(x) : h \in H,\ h \geq f\ \mu\text{-a.e.}\} = f(x) = \sup\{h(x) : h \in H,\ h \leq f\ \mu\text{-a.e.}\}$$

for μ-a.e. $x \in X$. Thus each of the following statements implies the next:

i) $M_x^{\emptyset}(H) = \{\varepsilon_x\}$ for μ-a.e. $x \in X$.

ii) $A_S(H)$ is $\mathscr{L}^p$-dense in $C(X)$.

iii) $A(H)$ is dense in $L^p(\mu)$.

Example 7.32 below shows that, just as in the $C_o(X)$-case, even condition (iii) is not necessary for H to be a Korovkin space in $L^p(\mu)$. As an auxiliary result we need the following slight modification of Counterexample 7.10:

<u>7.31 Lemma</u>: Let μ be a positive Radon measure on the two-point compactification $\bar{\mathbb{R}} := \mathbb{R} \cup \{-\infty,\infty\}$ of $\mathbb{R}$ such that $\{-\infty,\infty\} \cup [-1,1]$ is contained in the support of μ and $\mu(\{-\infty,\infty\}) = 0$. If $g_1,\ldots,g_5$ are the functions defined in Example 7.10 set

$$h_i(x) = \frac{g_i(x)}{1+|x|^3} \qquad (x \in \mathbb{R})$$

for $i = 1,\ldots,5$, and extend h_i by continuity to $\bar{\mathbb{R}}$. Given $p \in [1,\infty[$ the equivalence classes $h_1,\ldots,h_5$ of $h_1,\ldots,h_5$ form a Korovkin system in $L^p(\mu)$, but the H-affine elements are not dense in $L^p(\mu)$, where H is the linear hull of $h_1,\ldots,h_5$.

<u>Proof</u>: Let H be the linear space generated by $h_1,\ldots,h_5$. Since $N := \{-\infty,\infty\}$ is μ-negligible, the first part of the assertion will follow from the equality

$$M_x^N(H) = \{\varepsilon_x\} \qquad \text{for all } x \in \mathbb{R} = \bar{\mathbb{R}} \setminus N,$$

by Theorem 7.27. The proof of this equality is an obvious modification of the arguments used in Example 7.10. In order to show that $A(H)$ is not dense in $L^p(\mu)$, note first that the linear form $h \to h(\infty)$ defined on H has a linear extension φ_∞ to the space

$$\tilde{H} := \{ f \in \mathcal{L}^p(\mu) : \exists h_1, h_2 \in H \quad h_1 \le f \le h_2 \quad \mu\text{-a.e.}\}$$

dominated by the sublinear functional

$$\rho(f) = \inf\{ h(\infty) : h \in H,\ h \ge f \ \mu\text{-a.e.}\}.$$

Observe that $\varphi_\infty(f) \ge 0$ whenever $f \ge 0$ μ-a.e., since $\varphi_\infty(-f) \le \rho(-f) \le 0$. It follows that $\varphi_\infty(f_1) \le \varphi_\infty(f_2)$ for all $f_1, f_2 \in \tilde{H}$ satisfying $f_1 \le f_2$ μ-a.e. Similarly, there exist linear forms $\varphi_{-\infty}, \varphi_{-1}, \varphi_1$ on $\tilde{H}$ such that $\varphi_{-\infty}(h) = h(-\infty)$, $\varphi_{-1}(h) = h(-1)$, $\varphi_1(h) = h(1)$ for all $h \in H$ and $\varphi_i(f) \le \varphi_i(f_2)$ whenever $f_1 \le f_2$ μ-a.e., for $i = -\infty, -1, 1$. Given $x \in [-1,1]$, consider the linear form

$$\tau_x := \alpha_x^{(-\infty)} \cdot \varphi_{-\infty} + \alpha_x^{(-1)} \cdot \varphi_{-1} + \alpha_x^{(1)} \cdot \varphi_1 + \alpha_x^{(\infty)} \cdot \varphi_\infty \quad \text{on } \tilde{H},$$

where $\alpha_x^{(\infty)} = \dfrac{1}{4(1+|x|^3)}\left((1-x) + 27(1+x) - 2(x+2)^3\right)$, $\alpha_x^{(-\infty)} = 3\alpha_x^{(\infty)}$,

$\alpha_x^{(-1)} = \dfrac{1-x}{1+|x|^3}$, $\alpha_x^{(1)} = \dfrac{1+x}{1+|x|^3}$. It is easy to check that $\tau_x(h_i) = h_i(x)$

for $i = 1,\ldots,5$ and that $\alpha_x^{(-\infty)}$, $\alpha_x^{(-1)}$, $\alpha_x^{(1)}$, $\alpha_x^{(\infty)}$ are non-negative. If $f \in \mathcal{L}^p(\mu)$ belongs to an H-affine equivalence class $f \in L^p(\mu)$, we obtain

$$h(x) = \tau_x(h) \le \tau_x(f) \le \tau_x(h') = h'(x)$$

for $h, h' \in H$, $h \le f \le h'$ μ-a.e. It follows that

$$\sup\{h(x) : h \in H,\ h \le f \ \mu\text{-a.e.}\} \le \tau_x(f) \le \inf\{h'(x) : h' \in H,\ h' \ge f \ \mu\text{-a.e.}\},$$

or, since f is H-affine, $f(x) = \tau_x(f)$ for μ-a.e. $x \in [-1,1]$. Moreover, from the definition of τ_x we conclude that $x \to \tau_x(f)$, $x \in [-1,1]$, is the restriction of a function in H. Hence, there exists $h \in H$ such that $f|_{[-1,1]} = h|_{[-1,1]}$ μ-a.e. Since $[-1,1] \subset \operatorname{supp}(\mu)$, $A(H)$ is not dense in $L^p(\mu)$. ■

<u>7.32 Corollary</u>: Given $p \in [1,\infty[$, there exists a 5-dimensional linear subspace H of $C([-1,1])$ with the following properties:

i) $\mathcal{H}$ contains the constant functions and separates the points of $[-1,1]$.

ii) If λ is the Lebesgue measure on $[-1,1]$ and H is the linear subspace of $L^p(\lambda)$ corresponding to $\mathcal{H}$, then H is a Korovkin space in $L^p(\lambda)$.

iii) The H-affine elements are not dense in $L^p(\mu)$.

Proof: If $\Phi : [-1,1] \to \bar{\mathbb{R}}$ is given by $\Phi(x) = x/(1-|x|)$, then Φ is a homeomorphism of $[-1,1]$ onto $\bar{\mathbb{R}}$ and the image $\Phi(\mu)$ has support $\bar{\mathbb{R}}$. Hence, if $h_1,\ldots,h_5$ are the functions defined in Lemma 7.31, we deduce that the linear subspace H_Φ generated by the equivalence classes of $h_1 \circ \Phi,\ldots,h_5 \circ \Phi$ in $L^p(\lambda)$ form a Korovkin system in $L^p(\lambda)$ and that the H_Φ-affine elements are not dense in $L^p(\lambda)$. Moreover, the function $h_1 \circ \Phi$ is strictly positive on $]-1,1[$ and $(h_5 \circ \Phi)(-1) = (h_5 \circ \Phi)(1) = 1$. Therefore, $s := h_5 \circ \Phi + n(h_1 \circ \Phi)$ is a strictly positive function in the linear hull $\mathcal{H}_\Phi$ of $h_1 \circ \Phi,\ldots,h_5 \circ \Phi$ for sufficiently large $n \in \mathbb{N}$.
Let $\mathcal{H} := \{g/s : g \in \mathcal{H}_\Phi\}$. If H is the corresponding subspace of $L^p(\mu)$, an element $f \in L^p(\mu)$ is H-affine iff it is of the form $f = g/s$ for some H_Φ-affine element $g \in L^p(\lambda)$, which shows that $A(H)$ is not dense in $L^p(\lambda)$. Since condition (i) is trivially satisfied, it remains to prove (ii). Note that the positive linear mappings $M_{s^{-1}}$, $M_s : L^p(\lambda) \to L^p(\lambda)$ defined by $M_{s^{-1}}(f) = f/s$, $M_s(f) = s \cdot f$ are continuous isomorphisms from $L^p(\lambda)$ onto itself, which are inverse to each other. Thus, if $(T_i)_{i\in I}$ is an equicontinuous net of positive operators on $L^p(\lambda)$ such that $(T_i h)_{i\in I}$ converges to h in $L^p(\lambda)$ for each $h \in H$, then

$$\lim_{i\in I}(M_s \circ T_i \circ M_{s^{-1}})(g) = g \quad \text{in } L^p(\lambda) \quad \text{for each } g \in H_\Phi.$$

Consequently, $\lim_{i\in I}(M_s \circ T_i \circ M_{s^{-1}})(f) = f$ for each $f \in L^p(\lambda)$, which yields that $\lim_{i\in I} T_i f = f$ for each $f \in L^p(\lambda)$. Therefore H is a Korovkin space in $L^p(\lambda)$. ∎

7.33 Remarks:

a) If H is the subspace of $L^p(\mu)$ defined in Example 7.30, where $\text{supp}(\mu) = \mathbb{R}^2$, it can be shown that an element $f \in L^p(\mu)$ is H-affine iff the equivalence class f contains a function $\mathfrak{f} \in \mathcal{L}^p(\mu)$ with the following property:

For each $r > 0$ there exist $\alpha,\beta,\gamma \in \mathbb{R}$ such that

$\mathfrak{f}(\xi,\eta) = \alpha\xi + \beta\eta + \gamma$ whenever $\xi^2 + \eta^2 = r^2$ $((\xi,\eta) \in \mathbb{R}^2)$.

Hence $A(H)$ is not dense in $L^p(\lambda)$ in this example, too.

b) Reviewing Theorem 7.27 the reader may ask for characterizations of finite Korovkin systems in $L^p(\mu)$ when X is not second countable. Besides the obvious generalization that only the support of μ is assumed to be second countable, we cannot hope to obtain more general results for the Radon measure case. Indeed, it is shown in [73] that finite Korovkin systems cannot exist in non-separable L^p-spaces.

c) Analyzing the proof of Theorem 7.20 we observe that we in fact checked the equality

$$\hat{\mathfrak{f}}^b(x) = \sup\{\int \mathfrak{f}\,d\eta : \eta \in M_x^b(H)\} \quad \text{for all } x \in \{b < \infty\}.$$

Since, in the same way, also the equality

$$\check{\mathfrak{f}}^b(x) = \inf\{\int \mathfrak{f}\,d\eta : \eta \in M_x^b(H)\}$$

holds for all $x \in \{b < \infty\}$, Theorem 7.18 remains true if we replace the set $P_x^b(H)$ by $M_x^b(H)$.

8. Convergence to vector lattice homomorphisms and essential sets

Although theorems of Korovkin type are most frequently concerned with sequences or nets of operators converging to the identity operator on some Banach lattice, there are several applications where the identity mapping is replaced by imbedding or restriction operators or, more generally, by vector lattice homomorphisms. In this case, we may consult the results obtained in section 6 to characterize Korovkin closures and Korovkin systems. For practical use, however, we would like to have descriptions that are easier to apply. We have learned in the last section how to use the envelopes $\check{f}$, $\hat{f}$ and $\check{\mathfrak{f}}^a, \hat{\mathfrak{f}}^a$, respectively, together with H-representing measures for efficient characterizations of Korovkin closures with respect to the identity operator. We are therefore looking for a description of $\mathrm{Kor}_{P_e,S}(H)$ for vector lattice homomorphisms S that is also based on these envelopes. It turns out that, in many cases, we need indeed only compare the upper and lower envelopes on certain sets determined by S called S-essential sets. If S is defined on $C_o(X)$, X locally compact, the support of S

$$\mathrm{supp}(S) = \bigcap_{g \in S^{-1}(\{0\})} g^{-1}(\{0\})$$

is, e.g., a S-essential set. In the same way, however, as the support of a positive Radon measure μ has to be refined sometimes by the system of all μ-measurable sets supported by μ, we shall obtain a description of $\mathrm{Kor}_{P_e,S}(H)$ only when we are using the complete system of S-essential sets.

In this section the vector lattice homomorphism S will most frequently be defined on a space of continuous real-valued functions. The corresponding result for L^p-spaces will be stated without proof, since it easily follows from 6.15, 7.13, 7.33, c, and the fact, that the kernel of a vector lattice homomorphism in L^p-spaces, $p \in [1,\infty[$ is a band (see [66], Ch. II. Cor. of 2.6 and 5.14):

8.1 Theorem: Given a locally compact, σ-compact space X, let μ be a positive Radon measure on X and let $(\Omega, \mathcal{A}, \nu)$ be an arbitrary measure space. Furthermore, let $S : L^p(\mu) \to L^q(\nu)$ be a vector lattice homomorphism for $p,q \in [1,\infty[$, $q \leq p$. Since the kernel $S^{-1}(\{0\})$ is a band in $L^p(\mu)$, there exists a μ-measurable subset $A \subset X$ such that

$$Se = 0 \Leftrightarrow |e| \wedge 1_A = 0 \qquad (e \in L^p(\mu),$$

where 1_A denotes the equivalence class of the characteristic function of A. If $\mathcal{H}$ is a finite-dimensional linear subspace of $\mathcal{L}^p(\mu)$ and H is the corresponding subspace of $L^p(\mu)$ then the following statements are equivalent for each $f \in \mathcal{L}^p(\mu)$:

i) The equivalence class f of f is in $\mathrm{Kor}_{P_e,S}(H)$.

ii) There exists a pth power μ-integrable numerical function $a : X \to]0,\infty]$ such that

$$\check{f}^a(x) = f(x) = \hat{f}^a(x) \qquad \text{for } \mu\text{-a.e. } x \in A .$$

iii) There is a μ-negligible set $N \subset X$ such that

$$\rho(f) = f(x) \text{ for all } x \in A \setminus N \text{ and all } \rho \in M_x^N(\mathcal{H}).$$

Finally, if the topology of X is second countable, then H is a Korovkin space with respect to P_e and S iff there exists a μ-negligible set $N \subset X$ such that

$$M_x^N(\mathcal{H}) = \{\varepsilon_x\} \qquad \text{for all } x \in A \setminus N.$$

In the rest of this section the lattice homomorphism S will be defined on a space $C_o(X)$, where X is an arbitrary locally compact space. As before we shall denote by H a (not necessarily finite dimensional) linear subspace of $C_o(X)$ and by G a Banach lattice possessing the bounded positive approximation property for G".

8.2 Definitions:

a) A subset $Y \subset X$ is called essential with respect to $S : C_o(X) \to G$ (or S-essential for short), if, for each downward directed subset $F \subset C_o(X)_+$ satisfying $\inf_{f \in F} f(x) = 0$ for all $x \in Y$, the net $(Sf)_{f \in F}$ converges to 0.

b) If we define $\hat{f}(x)$ and $\check{f}(x)$ for each $f \in C_o(X)$, $x \in X$ in the same way as in 7.1, i.e.,

$$\hat{f}(x) = \sup_{\varepsilon>0} \inf_{\substack{h \in H \\ h \geq f-\varepsilon}} h(x) \quad \text{and} \quad \check{f}(x) = \inf_{\varepsilon>0} \sup_{\substack{h \in H \\ h \leq f+\varepsilon}} h(x) ,$$

then the set

$$B_f := \{x \in X : \check{f}(x) = \hat{f}(x)\}$$

is called the bordering set of f.+)

8.3 Theorem: For each $f \in C_o(X)$ the following statements are equivalent

i) $f \in \mathrm{Kor}_{P_e,S}(H)$.

ii) $\lim_{k \in \hat{H}_{f,\varepsilon}} S(k \vee f) = Sf = \lim_{k \in \check{H}^{f,\varepsilon}} S(k \vee f)$ for each $\varepsilon > 0$.

iii) B_f is S-essential.

Proof: (i) $\Rightarrow$ (ii): Given $f \in \mathrm{Kor}_{P_e,S}(H)$ there exists a non-negative function $a \in C_o(X)$ such that

$$\lim_{k \in \hat{H}_{f-\varepsilon a}} S(k \vee f) = Sf = \lim_{k \in \check{H}^{f+\varepsilon a}} S(k \wedge f) \quad \text{for each } \varepsilon > 0,$$

by Theorem 6.17. Note that, for $\alpha := \max(1, \|a\|)$, $\hat{H}_{f-\varepsilon a} \subset \hat{H}_{f,\varepsilon\alpha}$ and

+) Usually the bordering set of f is the set $\{x \in X : f(x) = \hat{f}(x)\}$ (cf. [1],[18]). For our purpose, however, we prefer the definition given above.

$\check{H}^{f+\varepsilon a} \subset \check{H}^{f,\varepsilon\alpha}$, respectively. Moreover, $\hat{H}_{f,\varepsilon\alpha}$ is downward directed, while $\check{H}^{f,\varepsilon\alpha}$ is upward directed. This yields (ii).

(ii) ⇒ (i): If $\lim\limits_{k\in\hat{H}_{f,\varepsilon}} S(k \vee f) = Sf = \lim\limits_{k\in\check{H}^{f,\varepsilon}} S(k \wedge f)$ for each $\varepsilon > 0$,

then

$$\inf_{k\in\hat{H}_{f,\varepsilon}} \ell(Sk) \leq \inf_{k\in\hat{H}_{f,\varepsilon}} \ell(S(k \vee f)) = \ell(Sf) = \sup_{k\in\check{H}^{f,\varepsilon}} \ell(S(k \wedge f)) \leq \sup_{k\in\check{H}^{f,\varepsilon}} \ell(Sk),$$

for each $\varepsilon > 0$, $\ell \in G'_+$. Consequently,

$$\hat{S}_H(f)(\ell) = \sup_{\varepsilon>0} \inf_{k\in\hat{H}_{f,\varepsilon}} \ell(Sk) \leq \ell(Sf) \leq \inf_{\varepsilon>0} \sup_{k\in\check{H}^{f,\varepsilon}} \ell(Sk) = \check{S}_H(f)(\ell),$$

for each $\ell \in G'_+$. Since always $\check{S}_H f \leq \hat{S}_H f$, we conclude $f \in \mathrm{Kor}_{p_e,S}(H)$, by Theorem 6.7.

(iii) ⇒ (ii): Let B_f be essential with respect to S. For each $x \in B_f$ we then obtain

$$\begin{aligned}
\sup_{\varepsilon>0} \inf_{k\in\hat{H}_{f,\varepsilon}} (k(x) \vee f(x)) &= (\sup_{\varepsilon>0} \inf_{k\in\hat{H}_{f,\varepsilon}} k(x)) \vee f(x) = \\
&= (\sup_{\varepsilon>0} \inf_{\substack{h\in H\\ h\geq f-\varepsilon}} h(x)) \vee f(x) = \hat{f}(x) \vee f(x) \\
&= f(x) = \check{f}(x) \wedge f(x) = (\inf_{\varepsilon>0} \sup_{\substack{h\in H\\ h\leq f+\varepsilon}} h(x)) \wedge f(x) \\
&= \inf_{\varepsilon>0} \sup_{k\in\check{H}^{f,\varepsilon}} (k(x) \wedge f(x)).
\end{aligned}$$

Hence, for all $\varepsilon > 0$,

$$\inf_{k\in\hat{H}_{f,\varepsilon}} (k \vee f)(x) = f(x) = \sup_{k\in\check{H}^{f,\varepsilon}} (k \wedge f)(x),$$

which yields $\inf\limits_{(k,k')\in\hat{H}_{f,\varepsilon}\times\check{H}^{f,\varepsilon}} (k \vee f - k' \wedge f)(x) = 0$.

Since $\{k \vee f - k' \wedge f : k \in \hat{H}_{f,\varepsilon}, k' \in \check{H}^{f,\varepsilon}\}$ is downward directed, it follows that

$$\lim_{(k,k')\in\hat{H}_{f,\varepsilon}\times\check{H}^{f,\varepsilon}} S(k \vee f - k' \wedge f) = 0$$

using the fact that B_f is S-essential. Therefore,

$$\lim_{k\in \check{H}^{f,\varepsilon}} S(k\wedge f) = Sf = \lim_{k\in \hat{H}_{f,\varepsilon}} S(k\vee f) \quad \text{for each } \varepsilon > 0.$$

If we set $F_n := \{f\vee k - f\wedge k' : k\in \hat{H}_{f,1/n},\ k'\in \check{H}^{f,1/n}\}$ for each $n\in\mathbb{N}$, then the remaining implication (ii) ⇒ (iii) is an immediate consequence of the following Lemma. ■

<u>8.4 Lemma</u>: Given a decreasing sequence $(F_n)_{n\in\mathbb{N}}$ in $C_o(X)_+$ let $u : X \to \mathbb{R}_\infty$ be defined by

$$u(x) = \sup_{n\in\mathbb{N}} \inf_{f\in F_n} f(x) \quad \text{for all } x\in X.$$

If 0 is a cluster point of $S(F_n)$ for each $n\in\mathbb{N}$, then the set $Y := \{x\in X : u(x) = 0\}$ is essential with respect to S.

<u>Proof</u>: Let $F \subset C_o(X)_+$ be a downward directed set such that $\inf_{f\in F} f(x) = 0$ for all $x\in Y$. In order to show that $\lim_{f\in F} Sf = 0$, let $\mathcal{U}$ denote the system of all zero-neighborhoods in G. Given $U\in\mathcal{U}$ choose a symmetric solid neighborhood U' of 0 in G such that $U'+U'\subset U$. We define the sequences (f_n') in $C_o(X)_+$ and (V_n) in $\mathcal{U}$ inductively as follows:

Select $V_1\in\mathcal{U}$ and $f_1'\in F_1$ such that $V_1+V_1\subset U'$ and $Sf_1'\in V_1$. If f_m', V_m are already constructed for $m\in\mathbb{N}$ let $V_{m+1}\in\mathcal{U}$ and $f_{m+1}'\in F_{m+1}$ be such that $V_{m+1}+V_{m+1}\subset V_m$ and $Sf_{m+1}'\in V_{m+1}$. Setting $A := \{\sum_{i=1}^{\ell} f_i' : \ell\in\mathbb{N}\}$ the algebraic difference $F-A$ is downward directed, since the functions f_i' are non-negative. Therefore, $(F-A)^+ := \{k^+ \in F-A\}$ is downward directed, too.

We claim that $\inf_{k\in(F-A)^+} k(x) = 0$ for all $x\in X$.

This equality being evident for $x\in Y$ we need only prove it for $x\in X\setminus Y$. In this case, however, we have $u(x) > 0$. Hence, if $f\in F$ is arbitrary, we can find $j,\ell\in\mathbb{N}$ such that

$$(\ell + 1) \cdot \inf_{f' \in F_j} f'(x) \geq f(x).$$

Observing the inequality $f_n'(x) \geq \inf_{f' \in F_n} f'(x) \geq \inf_{f' \in F_j} f'(x)$ valid for all $n \geq j$, $n \in \mathbb{N}$, it follows that

$$\sum_{n=j}^{j+\ell} f_n'(x) \geq (\ell + 1) \cdot \inf_{f' \in F_j} f'(x) \geq f(x).$$

Consequently, $(f - \sum_{n=1}^{j+\ell} f_n')^+(x) = 0$, which yields $\inf_{k \in (F-A)^+} k(x) = 0$.

By Dini's theorem, there are functions $f_o \in F$ and $a \in A$ such that $S((f-a)^+) \in U'$ for all $f \in F$, $f \leq f_o$. On the other hand, we deduce from the definitions of A and of $(V_n)_{n \in \mathbb{N}}$ that $Sa \in U' = -U'$. Moreover, note that the inequality $-Sa \leq S(f-a) \leq S((f-a)^+)$ holds for all $f \leq f_o$, $f \in F$. Since U' is solid, we obtain $S(f-a) \in U'$ and $Sf = S(f-a) + Sa \in U' + U' \subset U$ for all $f \leq f_o$, $f \in F$. The neighborhood $U \in \mathcal{U}$ being arbitrary, we conclude that $\lim_{f \in F} Sf = 0$. ∎

Theorem 8.3 obviously simplifies the determination of $\mathrm{Kor}_{\mathcal{P}_e,S}(H)$ using S-essential sets. In fact, we have reduced the characterization of $\mathrm{Kor}_{\mathcal{P}_e,S}(H)$ to two separate easier problems:

1. Find out the upper and lower envelopes $\hat{f}$ and $\check{f}$.
2. Describe the essential sets with respect to S.

Since, for the solution of the first problem, we may use the results developed in the last section and, in particular, the technique of H-representing measures (see Lemma 7.5 and [6],[8],[9]), we shall now be mainly concerned with the second problem.

8.5 Examples of essential sets

a) Since Dini's theorem holds in $C_o(X)$, the whole set X is S-essential for every vector lattice homomorphism $S : C_o(X) \to G$.

b) If $Y \subset X$ is S-essential, then every set $Z \subset X$, $Z \supset Y$ is also S-essential.

c) With respect to the identity operator on $C_o(X)$ X is the only essential set. Suppose, to the contrary, that there exists an essential set $Y \subset X$, $Y \neq X$. Choose $x \in X \setminus Y$ and a downward directed set $F \subset C_o(X)_+$ such that $f(x) = 1$ for all $f \in F$ and $\inf_{f \in F} f(y) = 0$ for all $y \in Y$. Then $\lim_{f \in F} f = 0$, since Y is essential with respect to the identity mapping, contradicting the equality $f(x) = 1$ valid for all $f \in F$.

d) Given a finite positive Radon measure μ on X let $S : C_o(X) \to L^p(\mu)$, $1 \leq p < \infty$, be the natural imbedding operator. A subset $Y \subset X$ is S-essential iff $X \setminus Y$ has inner measure 0. To prove this let K be a compact subset of $X \setminus Y$, where Y is S-essential. The characteristic function 1_K of K being upper semi-continuous, there exists a downward directed set $F \subset C_o(X)_+$ of continuous functions with compact support such that

$$\inf_{f \in F} f(x) = 1_K(x) \quad \text{for all } x \in X.$$

Since Y is S-essential we conclude $\lim_{f \in F} Sf = 0$, i.e.

$$\mu(K) = \inf_{f \in F} \mu(f) = 0.$$

Conversely, consider a subset $Y \subset X$ such that $X \setminus Y$ has inner measure 0. If $A \subset C_o(X)_+$ is downward directed and $\inf_{f \in A} f(y) = 0$ for all $y \in Y$, then $f_o := \inf_{f \in A} f$ is upper semi-continuous and the μ-measurable subset $\{x \in X : f_o(x) \neq 0\}$ of $X \setminus Y$ has measure 0. Hence $\lim_{f \in A} Sf = 0$ (cf. [17], IV, § 4, no 4, Cor. 2), which shows that Y is S-essential.

<u>8.6 Application (cf. [11])</u>: Given a compact topological space K let H be a linear subspace of $C(K)$ containing a strictly positive function h_o. If μ is a positive Radon measure on K and $S : C(K) \to L^p(\mu)$, $1 \leq p < \infty$, denotes the natural imbedding operator, then the following statements are equivalent for each $f \in C(K)$:

i) $f \in \mathrm{Kor}_{P_e,S}(H)$

ii) $\inf_{h\in H_f} h(x) = f(x) = \sup_{h\in H^f} h(x)$ for μ-a.e. $x \in K$.

Proof: Since the order unit norm defined by h_o is equivalent to the sup-norm on $C(K)$, we obtain

$$\hat{f}(x) = \sup_{\varepsilon>0} \inf_{h\in H_{f-\varepsilon h_o}} h(x) = \lim_{\varepsilon\to 0} \inf_{h\in H_{f-\varepsilon h_o}} h(x) \qquad (\text{in } \mathbb{R}_\infty)$$

$$= \lim_{\varepsilon\to 0} \inf_{h\in H_{f-\varepsilon h_o}} h(x) + \varepsilon h_o(x) = \inf_{h'\in H_f} h'(x)$$

for all $x \in K$. Similarly, $\check{f}(x) = \sup_{h\in H^f} h(x)$ for all $x \in K$.

Thus, the assertion follows from Theorem 8.3 and Example 8.5, d. ∎

If there exists a smallest S-essential set we at once know all S-essential subsets by 8.5, b. As example 8.5, d indicates, however, there is no smallest S-essential set, in general. Nevertheless, we can prove that a smallest essential set with respect to the vector lattice homomorphism $S : C_o(X) \to G$ exists, whenever G is a Dini lattice (see [62]). We shall only prove this for Dini lattices which are Banach lattices. As standard examples the reader should keep in mind spaces of type $C_o(Y)$, Y locally compact, or ℓ^p, $p \in [1,\infty[$.

8.7 Theorem: Let V denote the set of all real-valued vector lattice homomorphisms on G. If $\lim_{i\in I} g_i = 0$ for each decreasing net $(g_i)_{i\in I}$ in G_+ such that $\inf_{i\in I} \ell(g_i) = 0$ for all $\ell \in V$, then the set

$$Y_o := \{x \in X : \varepsilon_x \in S'(V)\}$$

is the smallest S-essential subset of X. Here ε_x denotes the evaluation functional at x and S' is the adjoint of S.

Proof: Recall that each real vector lattice homomorphism on $C_o(X)$ can be written in the form $\alpha\varepsilon_x$, where $\alpha \in \mathbb{R}_+$, $x \in X$. Hence, for each $\ell \in V$ there exists $\alpha \in \mathbb{R}_+$ and $x \in Y_o$ such that $\alpha\varepsilon_x = \ell \circ S$. Let F be a downward directed subset of $C_o(X)_+$ satisfying

$$\inf_{f \in F} f(x) = 0 \qquad \text{for all } x \in Y_o.$$

Then $\inf_{f\in F} \ell(Sf) = 0$ for all $\ell \in V$, which implies that $\lim_{f\in F} Sf = 0$.
Consequently, Y_o is S-essential.
Suppose that we could find an essential set $Y \subset Y_o$ with respect to S, $Y \neq Y_o$. Choose $x_o \in Y_o \setminus Y$ and put $F := \{f \in C_o(X)_+ : f(x_o) = 1\}$. Then F is downward directed and $\inf_{f\in F} f(x) = 0$ for all $x \in Y$. Since Y is S-essential, we obtain $\lim_{f\in F} Sf = 0$. On the other hand, there exists $\ell \in V$ such that $\varepsilon_{x_o} = \ell \circ S$. This yields $\lim_{f\in F} \ell(Sf) = 1$ contradicting the continuity of ℓ. ■

8.8 Example: Given compact spaces X,Z let $S : C(X) \to C(Z)$ be a vector lattice homomorphism. If $Z_o := \{z \in Z : S(1)(z) \neq 0\}$, there exists a continuous function $\Phi : Z_o \to X$ such that

$$Sf(z) = \begin{cases} S1(z)\; f(\Phi(z)) & \text{for } z \in Z_o \\ 0 & \text{for } z \in Z\setminus Z_o \end{cases} \qquad (f \in C(X))$$

(see [61],[24],[15]). By Theorem 8.7, $\Phi(Z_o)$ is the smallest S-essential set. Hence, for a linear subspace $H \subset C(X)$, a function $f \in C(X)$ belongs to $\mathrm{Kor}_{P_e,S}(H)$ iff $\Phi(Z_o) \subset B_f$.

8.9 Theorem: Let H be an arbitrary linear subspace of $C_o(X)$, X locally compact. If the Choquet boundary $\partial_H(X)$ is S-essential, H is a Korovkin space with respect to S. The converse is also true, provided that X possesses a countable basis of open sets.

Proof: If $\partial_H(X)$ is essential with respect to S, then the bordering sets B_f are essential for all $f \in C_o(X)$, since, by Lemma 7.5,
$\partial_H(X) = \bigcap_{f\in C_o(X)} B_f$.
From Theorem 8.3 we deduce that H is a Korovkin space.
Conversely, let H be a Korovkin space in $C_o(X)$ with respect to S and suppose X to be second countable. Since $C_o(X)$ is separable, there exists a dense sequence (f_n) in $C_o(X)$. For each $n \in \mathbb{N}$, B_{f_n} is S-essential by Theorem 8.3. Moreover, $\bigcap_{n\in\mathbb{N}} B_{f_n} \subset B_f$ for each $f \in C_o(X)$. Indeed, if we select a subsequence $(f_{n_k})_{k\in\mathbb{N}}$ from $(f_n)_{n\in\mathbb{N}}$ converging to f, then for every H - representing measure $\mu \in M_x(H)$, $x \in \bigcap_{n\in\mathbb{N}} B_{f_n} \subset \bigcap_{k\in\mathbb{N}} B_{f_{n_k}}$ we obtain $\mu(f) = \lim_{k\to\infty} \mu(f_{n_k}) = \lim_{k\to\infty} f_{n_k}(x) = f(x)$.
Therefore, $x \in B_f$. From this we conclude that $\partial_H(X) = \bigcap_{f\in C_o(X)} B_f = \bigcap_{n\in\mathbb{N}} B_{f_n}$. Hence the following lemma will complete the proof:

8.10 Lemma: Let $(f_n)_{n\in\mathbb{N}}$ be a sequence in $\mathrm{Kor}_{P_e,S}(H)$. Then $\bigcap_{n\in\mathbb{N}} B_{f_n}$ is S-essential.

Proof: Let $\mathcal{U}$ denote the system of all zero-neighborhoods in G. For each $U \in \mathcal{U}$ we can choose a sequence $(V_n^U)_{n\in\mathbb{N}}$ in $\mathcal{U}$ satisfying $V^U + V_1^U \subset U$ and $V_{n+1}^U + V_{n+1}^U \subset V_n^U$ for all $n \in \mathbb{N}$. Given $\varepsilon > 0$ and $n \in \mathbb{N}$ select $f_n^{U,\varepsilon} \in \{k \vee f_n - k' \wedge f_n : k \in \hat{H}_{f_n,\varepsilon},\ k' \in \check{H}^{f_n,\varepsilon}\}$ such that $S(f_n^{U,\varepsilon}) \in V_n^U$. This is possible, since

$$\lim_{(k,k')\in\hat{H}_{f_n,\varepsilon}\times\check{H}^{f_n,\varepsilon}} S(k \vee f_n - k' \wedge f_n) = 0.$$

Setting $f_{U,n} := \sum_{i=1}^{n} f_i^{U,1/n}$, we obtain $S(f_{U,n}) \in \sum_{i=1}^{n} V_i^U \subset U$ for each $n \in \mathbb{N}$, $U \in \mathcal{U}$. The sets $F_n := \{f_{U,m} : m \geq n,\ U \in \mathcal{U}\}$ form a decreasing sequence. Furthermore, 0 is a cluster point of $S(F_n)$ for each $n \in \mathbb{N}$, since $\lim_{U \in \mathcal{U}} S(f_{U,n}) = 0$. By Lemma 8.4 the set

$$\{x \in X : \sup_{n \in \mathbb{N}} \inf_{f \in F_n} f(x) = 0\}$$

is S-essential. To complete the proof it therefore suffices to show that this set is contained in $\bigcap_{n \in \mathbb{N}} B_{f_n}$, or, equivalently, that

$$\sup_{n \in \mathbb{N}} \inf_{f \in F_n} f(x) > 0 \quad \text{for all } x \in X \setminus \bigcap_{n \in \mathbb{N}} B_{f_n}.$$

Given $x \in X \setminus \bigcap_{n \in \mathbb{N}} B_{f_n}$ choose $n \in \mathbb{N}$ such that $x \notin B_{f_n}$. Then $0 < \rho < \hat{f}_n(x) - \check{f}_n(x)$ for some $\rho \in \mathbb{R}$. Observing the equality

$$\hat{f}_n(x) - \check{f}_n(x) = \sup_{\varepsilon > 0} \inf_{k \in \hat{H}_{f_n,\varepsilon}} (k(x) \vee f_n(x)) - \inf_{\varepsilon > 0} \sup_{k' \in \check{H}^{f_n,\varepsilon}} (k'(x) \wedge f_n(x))$$

$$= \sup_{\varepsilon > 0} \inf_{k \in \hat{H}_{f_n,\varepsilon},\, k' \in \check{H}^{f_n,\varepsilon}} (k \vee f_n - k' \wedge f_n)(x)$$

we can find $m \in \mathbb{N}$, $m \geq n$, such that

$$\rho < \inf_{k \in \hat{H}_{f_n,\varepsilon},\, k' \in \check{H}^{f_n,\varepsilon}} (k \vee f_n - k' \wedge f_n)(x) \quad \text{for all } \varepsilon \in]0, \tfrac{1}{m}].$$

Consequently, $f_n^{U,1/j}(x) > \rho$ and $f_{U,j}(x) > \rho$ for each $U \in \mathcal{U}$, $j \geq m$ ($j \in \mathbb{N}$). It follows that $\inf_{f \in F_m} f(x) \geq \rho$, hence also

$$\sup_{j \in \mathbb{N}} \inf_{f \in F_j} f(x) \geq \rho > 0. \qquad \blacksquare$$

<u>Remark</u>: In this section the exclusion of general locally convex vector lattices has become an inconvenient restriction. Indeed, spaces like $C(X)$, X locally compact endowed with the topology of uniform convergence on compacts and many other practically useful spaces may replace $C_o(X)$.

Hence the proofs concerning S-essential sets have been organized so that they remain valid for arbitrary Dini lattices instead of $C_o(X)$.

List of symbols

Notation	meaning
$\bar{A}$	(topological) closure of A
$f\|_K$	restriction of f
$\mathbb{R}_+$	$\{x \in \mathbb{R} : x \geq 0\}$
$\mathbb{R}_+^*$	$\{x \in \mathbb{R} : x > 0\}$
$\mathbb{R}_\infty$	$\mathbb{R} \cup \{\infty\}$
x^+	positive part
x^-	negative part
$\wedge$, $\vee$	infimum, supremum (of two elements)
$\bigwedge$, $\bigvee$	infimum, supremum (of a family)
C_p	see page 4
$\mathscr{L}^p(\mu)$	see page 4
$L^p(\mu)$	see page 4
F_s	sup-completion of F, see page 10
ℓ^p	space of all real-valued sequences (ξ_n) such that $\sum_n \|\xi_n\|^p < \infty$
l.s.c.	lower semi-continuous
u.s.c.	upper semi-continuous
$\mathbf{K}_{p,x}$	see page 17
p_T	see page 17
$p^\cap$	regularization of p, see page 21
F_a	vector lattice ideal generated by a, see p. 42
$\tilde{p}^\otimes$	see page 45
r, $r^\otimes$	see page 52
$\hat{T}$	see page 56
F^b	order dual of F
F'	topological dual of F

notation	meaning
F^*	algebraic dual of F
O_ε	see page 47
$O_{\varepsilon,f}$	see page 48
O_ε^+	see page 53
$O_{\varepsilon,f}^+$	see page 54
$H_{e,\varepsilon}$	see page 55
$H_{e,f,\varepsilon}^\otimes$	see page 63
f_g	see page 84
$f_{i,\lambda}$	see page 86
$\mathcal{P}$, $\mathcal{P}_e$, $\mathcal{P}_o$, $\mathcal{P}_o''$	see page 105
$\mathrm{Kor}_{\mathcal{D},S}(H)$	see page 105
$\hat{H}_{e,\varepsilon}$, $\check{H}^{e,\varepsilon}$	see page 106
$\hat{S}_H$, $\check{S}_H$	see page 106
PBAP	positive bounded approximation property, see page 111
$\sigma(E,F)$	weak topology on E with respect to the dual pair (E,F)
$\lhd$	see page 120
$\hat{S}_H^a$, $\check{S}_H^a$	see page 120, 121
H_x	$\{h \in H : h \geq x\}$
H^x	$\{h \in H : h \leq x\}$
$\hat{H}_x$, $\check{H}^x$	see page 123
$\hat{f}$, $\check{f}$	see page 127
$\mathcal{M}_x(H)$	see page 129
$\mathcal{A}_S(H)$	see page 133
$\mathcal{A}(H)$	see page 134
$\mathcal{P}_x^b(H)$	see page 142
$\mathcal{L}_a$	see page 139
$\mathcal{H}_g$, $\mathcal{H}^g$	see page 139
$\{a = \infty\}$, $\{a < \infty\}$	$\{x : a(x) = \infty\}$, $\{x : a(x) < \infty\}$
$\hat{\mathfrak{s}}^a$, $\check{\mathfrak{s}}^a$	see page 140

notation	meaning
$M_x^a(H)$	see page 145
$K(X)$	continuous functions with compact support
$M_x^N(H)$	see page 150
B_f	bordering set of f, see page 164

References

1. ALFSEN, E.M.: Compact Convex Sets and Boundary Integrals. Springer-Verlag, Berlin - Heidelberg - New York 1971.

2. ANGER, B. and LEMBCKE, J.: Hahn-Banach type theorems for hypolinear functionals. Math. Ann. 209, 127-151 (1974).

3. ANGER, B. and LEMBCKE, J.: Extension of linear forms with strict domination on locally compact cones. Math.Scand. 47 (1980), 251-260.

4. ANDO, T.: Contractive projections in L_p-spaces. Pacific J. Math. 17, 391-405 (1966).

5. ANDO, T.: Banachverbände und positive Projektionen. Math. Z. 109, 121-130 (1969).

6. BAUER, H.: Theorems of Korovkin type for adapted spaces. Ann. Inst. Fourier 23, Fasc. 4, 245-260 (1973).

7. BAUER, H.: Wahrscheinlichkeitstheorie und Grundzüge der Maßtheorie. 2. Aufl. W. de Gruyter, Berlin 1974.

8. BAUER, H.: Convergence of monotone operators. Math. Z. 136, 315-330 (1974).

9. BAUER, H. and DONNER, K.: Korovkin approximation in $C_o(X)$. Math. Ann. 236, 225-237 (1978).

10. BAUER, H., LEHA, G. and PAPADOPOULOU, S.: Determination of Korovkin closures. Math. Z. 168, 263-274 (1979).

11. BERENS, H. and LORENTZ, G.G.: Theorems of Korovkin type for positive linear operators on Banach lattices. In: Approximation Theory. Ed. G.G. Lorentz. Academic Press, New York - London 1973. pp.1-30.

12. BERENS, H. and LORENTZ, G.G.: Sequences of contractions on L^1-spaces. J. Functional Analysis 15, 155-165 (1974).

13. BERENS, H. and LORENTZ, G.G.: Korovkin theorems for sequences of contractions on L^p-spaces. In: Linear Operators and Approx. II. Birkhäuser, Basel - Stuttgart 1974. 367-375.

14. BERENS, H. and LORENTZ, G.G.: Geometric theory of Korovkin sets. J. Approximation Theory 15, 161-189 (1975).

15. BONSALL, F.F.; LINDENSTRAUSS, J. and PHELPS, R.R.: Extreme positive operators on algebras of functions. Math. Scand. 18, 161-182 (1966).

16. BOURBAKI, N.: General Topology I. Hermann - Addison Wesley, Paris-Reading 1966.

17. BOURBAKI, N.: Intégration. Ch. 1-4. 2e édition. Hermann, Paris 1965.

18. CHOQUET, G.: Lectures on Analysis II. Math. Lecture Notes Series. Benjamin, Amsterdam - New York 1969.

19. DEVORE, R.A.: The approximation of continuous functions by positive linear operators. Lecture Notes in Math. 293. Springer-Verlag, Berlin - Heidelberg - New York 1972.

20. DONNER, K.: Korovkin theorems for positive linear operators. J. Approximation Theory 13, 443-450 (1975).

21. DONNER, K.: Korovkin closures for positive linear operators. J. Approximation Theory, 14-25 (1979).

22. DONNER, K.: Korovkin Theorems in L^p-spaces. J. Functional Analysis 42, 12-28 (1981).

23. DZJADYK, V.K.: Approximation of functions by positive linear operators and singular integrals (Russian). Math. Sbornik (N.S.) 70 (112), 508-517. MR 34 # 8053 (1966).

24. ELLIS, A.L.: Extreme positive operators. Quart. J. Math. Oxford Ser. 15, 342-344 (1964).

25. ENGMANN, H.: Notwendige und hinreichende Ungleichungen für die Existenz spezieller L^1-Kontraktionen. Z. Wahrscheinlichkeitstheorie verw. Gebiete 33, 317-329 (1976).

26. FAKHOURY, H.: Le théorème de Korovkin dans $C(X)$ et $L^p(\mu)$. Sém. Choquet, Initiation à l'Analyse, 13e année, n° 9, 20 p. (1973/74).

27. FEYEL, D.: Deux applications d'une extension du théorème de Hahn-Banach. C.R.Acad.Sci. Paris 280, 193-196 (1972).

28. FLÖSSER, H.O.: A Korovkin type theorem in locally convex M-spaces. Proc. Amer. Math. Soc. 72, 456-460 (1978).

29. FLÖSSER, H.O.; IRMISCH, R. and ROTH, W.: Infimum stable convex cones and approximation. Preprint Nr. 457. TH Darmstadt 1978. 25 p.

30. FLÖSSER, H.O.: Korovkin closures of finite sets. Arch. Math. 32, 600-608 (1979).

31. FLÖSSER, H.O. and ROTH, W.: Korovkin-Hüllen in Funktionenräumen. Math. Z. 166, 187-203 (1979).

32. HÖRMANDER, L.: Sur la fonction d'appui des ensembles convexes dans un espace localement convexe. Ark. Math. 3, 181-186 (1955).

33. IONESCU-TULCEA, A. and IONESCU-TULCEA, C.: Topics in the Theory of Lifting. Springer-Verlag, Berlin - Heidelberg - New York 1969.

34. JAMES, R.L.: The extension and convergence of positive operators. J. Approximation Theory 7, 186-197 (1973).

35. JAMES, R.L.: Korovkin sets in locally convex function spaces. J. Approximation Theory 12, 205-209 (1974).

36. JAMESON, G.: Ordered linear spaces. Lecture Notes in Math. 141. Springer-Verlag, Berlin - Heidelberg - New York 1970.

37. KATZNELSON, Y.: An Introduction to Harmonic Analysis. Wiley and Sons, New York - London - Sidney - Toronto 1968.

38. KITTO, W. and WULBERT, D.E.: Korovkin approximation in L_p-spaces Pacific J. Math. 63, 153-167 (1976).

39. KOROVKIN, P.P.: About the convergence of linear positive operators in the space of continuous functions. Dokladi Akad. Nauk SSSR 90, Nr. 6, 961-964 (1953).

40. KOROVKIN, P.P.: Linear Operators and Approximation Theory. Hind. Publ. Comp., Delhi 1960.

41. KRASNOSEL'SKII, M.A. and LIFŠIC, E.A.: A principle of convergence of sequences of positive linear operators. (Russian). Studia Math. 31, 455-468, MR 38 # 6372 (1968).

42. KÖNIG, H.: Sublineare Funktionale. Arch. Math. 23, 500-508 (1972).

43. LACEY, H.E.: The Isometric Theory of Classical Banach Lattices. Springer-Verlag, Berlin - Heidelberg - New York 1974.

44. LEVY, M.: Prolongement d'un opérateur d'un sous-espace de $L^1(\mu)$ dans $L^1(\nu)$. Sém. d'Analyse Fonctionnelle 1979-1980. École Polytechnique, Centre de Math. Palaiseau. Exp. No. V, Nov. 1979, V5 p.

45. LINDENSTRAUSS, J.: Extension of compact operators. Mem. Amer. Math. Soc. 48 (1964).

46. LINDENSTRAUSS, J.: On projections with norm 1 - an example. Proc. Amer. Math. Soc. 15, 403-406 (1964).

47. LINDENSTRAUSS, J.: Classical Banach Spaces. Lecture Notes in Math. 338. Springer-Verlag, Berlin - Heidelberg - New York 1973.

48. LORENTZ, G.G.: Positive and monotone approximation. In: Linear Operators and Approximation. Ed. P.L. Butzer, J.P. Kahane, B. Sz.-Nagy. Birkhäuser Verlag, Basel - Stuttgart 1971, pp. 284-291.

49. LOTZ, H.P.: Extensions and liftings of positive linear mappings on Banach lattices. Trans. Amer. Math. Soc. 211, 85-100 (1975).

50. LUSKY, W.: Some consequences of Rudin's paper "L^p-Isometries and equimeasurability". Indiana Univ. Math. J. 27, 859-866 (1978).

51. LUSKY, W.: Zur Geometrie von L_p-Räumen. Habilitationsschrift. Gesamthochschule Paderborn 1979.

52. LUXEMBURG, W.A.J.: Rearrangement-invariant Banach function spaces. Proc. Symp. Analysis. Queen's University Kingston, Ontario, 83-144, 1967.

53. LUXEMBURG, W.A.J. and ZAANEN, A.C.: Riesz Spaces I. North Holland, Amsterdam 1971.

54. MALIVERT, C.; PENOT, J.P. and THERA, M.: Un prolongement du théorème de Hahn-Banach. C.R.Acad.Sci. Paris 286, Sér.A, 165-168 (1978).

55. NACHBIN, L.: Some problems in extending and lifting linear transformations. Proc. Internat. Sympos. Linear Spaces. Jerusalem Acad. Press, Jerusalem. Pergamon Press, Oxford etc. 1961. pp. 340-350.

56. OLTENAU, O.: Convexité et prolongement d'opérateurs linéaires. C.R. Acad.Sci. Paris 286, Sér.A, 511-514 (1978).

57. PAPADOPOULOU, S.: Über den stationären Korovkin-Abschluß eines Funktionenraumes. Habilitationsschrift. Universität Erlangen-Nürnberg 1979.

58. PENOT, J.P. and THERA, M.: Conjugate vector-valued convex mappings. Preprint 1979.

59. PENOT, J.P. and THERA, M.: Semi-continuité des applications et des multiapplications. C.R.Acad.Sci. Paris 288, Sér.A, 241-244 (1979).

60. PERESSINI, A.L.: Ordered Topological Vector Spaces. Harper and Row, New York 1967.

61. PHELPS, R.R.: Extremal operators and homomorhisms. Trans. Amer. Math. Soc. 108, 265-274 (1963).

62. PORTENIER, C.: Caractérisation de certains espaces de Riesz. Sém. Choquet, Initiation à l'Analyse, 10e année, n° 6 (1970/71).

63. ŠAŠKIN, Yu. A.: Korovkin systems in spaces of continuous functions. Amer. Math. Soc. Transl. II, 54, 125-144 (1966).

64. ŠAŠKIN, Yu.A.: The Milman-Choquet boundary and approximation theory. Funct. Anal. Appl. 1, 170-171 (1967).

65. SCHAEFER, H.H.: Topological Vector Spaces. Springer-Verlag, Berlin - Heidelberg - New York 1971.

66. SCHAEFER, H.H.: Banach lattices and Positive Operators. Springer-Verlag, Berlin - Heidelberg - New York 1974.

67. SCHEFFOLD, E.: Über die punktweise Konvergenz von Operatoren in C(X). Rev. Acad. Ci. Zaragoza, II. Ser., 5-12 (1973).

68. SCHEFFOLD, E.: Ein allgemeiner Korovkin-Satz für lokalkonvexe Vektorverbände. Math. Z. 132, 209-214 (1973).

69. TZAFRIRI, L.: An isomorphic characterization of L_p- and c_o-spaces II. Michigan Math. J. 18, 21-31 (1971).

70. VASIL'EV, R.K.: The conditions for the convergence of isotonic operators in partially ordered sets with convergence classes. Math. Notes 12, 632-637 (1972).

71. VULIKH, B.Z.: Introduction to the Theory of Partially Ordered Spaces. Gordon and Breach, New York 1967.

72. WOLFF, M.: Über Korovkin-Sätze in lokalkonvexen Vektorverbänden. Math. Ann. 204, 49-56 (1973).

73. WOLFF, M.: Über die Korovkinhülle von Teilmengen in lokalkonvexen Vektorverbänden. Math. Ann. 213, 97-108 (1975).

74. WOLFF, M.: On the universal Korovkin closure of subsets in vector lattices. J. Approximation Theory 22, 243-253 (1978).

75. WOLFF, M.: On the theory of approximation by positive linear operators in vector lattices. In: Functional Analysis: Surveys and recent results. Ed. K.-D. Bierstedt, B. Fuchssteiner. North-Holland Math. Stud. 27, 73-87 (1977).

76. WOLFF, M.: Eine Bemerkung zum Desintegrationssatz von Strassen. Arch. Math. 28, 98-101 (1977).

77. WULBERT, D.E.: Convergence of operators and Korovkin's theorem. J. Approximation Theory 1, 381-390 (1968).

78. ZOWE, J.: Linear maps majorized by a sublinear map. Arch. Math. XXV, 637-645 (1975).

79. ZOWE, J.: Sandwich theorems for convex operators with values in an ordered vector space. J. Math. Anal. Appl. 66, 282-296 (1978).

Index

Vol. 817: L. Gerritzen, M. van der Put, Schottky Groups and Mumford Curves. VIII, 317 pages. 1980.

Vol. 818: S. Montgomery, Fixed Rings of Finite Automorphism Groups of Associative Rings. VII, 126 pages. 1980.

Vol. 819: Global Theory of Dynamical Systems. Proceedings, 1979. Edited by Z. Nitecki and C. Robinson. IX, 499 pages. 1980.

Vol. 820: W. Abikoff, The Real Analytic Theory of Teichmüller Space. VII, 144 pages. 1980.

Vol. 821: Statistique non Paramétrique Asymptotique. Proceedings, 1979. Edited by J.-P. Raoult. VII, 175 pages. 1980.

Vol. 822: Séminaire Pierre Lelong-Henri Skoda, (Analyse) Années 1978/79. Proceedings. Edited by P. Lelong et H. Skoda. VIII, 356 pages, 1980.

Vol. 823: J. Král, Integral Operators in Potential Theory. III, 171 pages. 1980.

Vol. 824: D. Frank Hsu, Cyclic Neofields and Combinatorial Designs. VI, 230 pages. 1980.

Vol. 825: Ring Theory, Antwerp 1980. Proceedings. Edited by F. van Oystaeyen. VII, 209 pages. 1980.

Vol. 826: Ph. G. Ciarlet et P. Rabier, Les Equations de von Kármán. VI, 181 pages. 1980.

Vol. 827: Ordinary and Partial Differential Equations. Proceedings, 1978. Edited by W. N. Everitt. XVI, 271 pages. 1980.

Vol. 828: Probability Theory on Vector Spaces II. Proceedings, 1979. Edited by A. Weron. XIII, 324 pages. 1980.

Vol. 829: Combinatorial Mathematics VII. Proceedings, 1979. Edited by R. W. Robinson et al.. X, 256 pages. 1980.

Vol. 830: J. A. Green, Polynomial Representations of GL_n. VI, 118 pages. 1980.

Vol. 831: Representation Theory I. Proceedings, 1979. Edited by V. Dlab and P. Gabriel. XIV, 373 pages. 1980.

Vol. 832: Representation Theory II. Proceedings, 1979. Edited by V. Dlab and P. Gabriel. XIV, 673 pages. 1980.

Vol. 833: Th. Jeulin, Semi-Martingales et Grossissement d'une Filtration. IX, 142 Seiten. 1980.

Vol. 834: Model Theory of Algebra and Arithmetic. Proceedings, 1979. Edited by L. Pacholski, J. Wierzejewski, and A. J. Wilkie. VI, 410 pages. 1980.

Vol. 835: H. Zieschang, E. Vogt and H.-D. Coldewey, Surfaces and Planar Discontinuous Groups. X, 334 pages. 1980.

Vol. 836: Differential Geometrical Methods in Mathematical Physics. Proceedings, 1979. Edited by P. L. García, A. Pérez-Rendón, and J. M. Souriau. XII, 538 pages. 1980.

Vol. 837: J. Meixner, F. W. Schäfke and G. Wolf, Mathieu Functions and Spheroidal Functions and their Mathematical Foundations Further Studies. VII, 126 pages. 1980.

Vol. 838: Global Differential Geometry and Global Analysis. Proceedings 1979. Edited by D. Ferus et al. XI, 299 pages. 1981.

Vol. 839: Cabal Seminar 77 - 79. Proceedings. Edited by A. S. Kechris, D. A. Martin and Y. N. Moschovakis. V, 274 pages. 1981.

Vol. 840: D. Henry, Geometric Theory of Semilinear Parabolic Equations. IV, 348 pages. 1981.

Vol. 841: A. Haraux, Nonlinear Evolution Equations- Global Behaviour of Solutions. XII, 313 pages. 1981.

Vol. 842: Séminaire Bourbaki vol. 1979/80. Exposés 543-560. IV, 317 pages. 1981.

Vol. 843: Functional Analysis, Holomorphy, and Approximation Theory. Proceedings. Edited by S. Machado. VI, 636 pages. 1981.

Vol. 844: Groupe de Brauer. Proceedings. Edited by M. Kervaire and M. Ojanguren. VII, 274 pages. 1981.

Vol. 845: A. Tannenbaum, Invariance and System Theory: Algebraic and Geometric Aspects. X, 161 pages. 1981.

Vol. 846: Ordinary and Partial Differential Equations, Proceedings. Edited by W. N. Everitt and B. D. Sleeman. XIV, 384 pages. 1981.

Vol. 847: U. Koschorke, Vector Fields and Other Vector Bundle Morphisms - A Singularity Approach. IV, 304 pages. 1981.

Vol. 848: Algebra, Carbondale 1980. Proceedings. Ed. by R. K. Amayo. VI, 298 pages. 1981.

Vol. 849: P. Major, Multiple Wiener-Itô Integrals. VII, 127 pages. 1981.

Vol. 850: Séminaire de Probabilités XV. 1979/80. Avec table générale des exposés de 1966/67 à 1978/79. Edited by J. Azéma and M. Yor. IV, 704 pages. 1981.

Vol. 851: Stochastic Integrals. Proceedings, 1980. Edited by D. Williams. IX, 540 pages. 1981.

Vol. 852: L. Schwartz, Geometry and Probability in Banach Spaces. X, 101 pages. 1981.

Vol. 853: N. Boboc, G. Bucur, A. Cornea, Order and Convexity in Potential Theory: H-Cones. IV, 286 pages. 1981.

Vol. 854: Algebraic K-Theory. Evanston 1980. Proceedings. Edited by E. M. Friedlander and M. R. Stein. V, 517 pages. 1981.

Vol. 855: Semigroups. Proceedings 1978. Edited by H. Jürgensen, M. Petrich and H. J. Weinert. V, 221 pages. 1981.

Vol. 856: R. Lascar, Propagation des Singularités des Solutions d'Equations Pseudo-Différentielles à Caractéristiques de Multiplicités Variables. VIII, 237 pages. 1981.

Vol. 857: M. Miyanishi. Non-complete Algebraic Surfaces. XVIII, 244 pages. 1981.

Vol. 858: E. A. Coddington, H. S. V. de Snoo: Regular Boundary Value Problems Associated with Pairs of Ordinary Differential Expressions. V, 225 pages. 1981.

Vol. 859: Logic Year 1979-80. Proceedings. Edited by M. Lerman, J. Schmerl and R. Soare. VIII, 326 pages. 1981.

Vol. 860: Probability in Banach Spaces III. Proceedings, 1980. Edited by A. Beck. VI, 329 pages. 1981.

Vol. 861: Analytical Methods in Probability Theory. Proceedings 1980. Edited by D. Dugué, E. Lukacs, V. K. Rohatgi. X, 183 pages. 1981.

Vol. 862: Algebraic Geometry. Proceedings 1980. Edited by A. Libgober and P. Wagreich. V, 281 pages. 1981.

Vol. 863: Processus Aléatoires à Deux Indices. Proceedings, 1980. Edited by H. Korezlioglu, G. Mazziotto and J. Szpirglas. V, 274 pages. 1981.

Vol. 864: Complex Analysis and Spectral Theory. Proceedings, 1979/80. Edited by V. P. Havin and N. K. Nikol'skii, VI, 480 pages. 1981.

Vol. 865: R. W. Bruggeman, Fourier Coefficients of Automorphic Forms. III, 201 pages. 1981.

Vol. 866: J.-M. Bismut, Mécanique Aléatoire. XVI, 563 pages. 1981.

Vol. 867: Séminaire d'Algèbre Paul Dubreil et Marie-Paule Malliavin. Proceedings, 1980. Edited by M.-P. Malliavin. V, 476 pages. 1981.

Vol. 868: Surfaces Algébriques. Proceedings 1976-78. Edited by J. Giraud, L. Illusie et M. Raynaud, V, 314 pages. 1981.

Vol. 869: A. V. Zelevinsky, Representations of Finite Classical Groups. IV, 184 pages. 1981.

Vol. 870: Shape Theory and Geometric Topology. Proceedings, 1981. Edited by S. Mardešić and J. Segal. V, 265 pages. 1981.

Vol. 871: Continuous Lattices. Proceedings, 1979. Edited by B. Banaschewski and R.-E. Hoffmann. X, 413 pages. 1981.

Vol. 872: Set Theory and Model Theory. Proceedings, 1979. Edited by R. B. Jensen and A. Prestel. V, 174 pages. 1981.